COMBAT LEGEND

F-86 SABRE

Martin W. Bowman

Airlife

Acknowledgements

The author would like to thank the following for their help in producing this book:

Mrs Dorothy Adams; Larry Davis; Merle and Grp Capt. R. J. F. Dickinson, AFC, DFC (US), RAF Retd.; GMS Enterprises; Eric Haywood; Harry Holmes; Grp Capt. Mohammed Shaukat-ul-Islam, PAF Retd.; Robert Jackson; David Ketteringham; Paul Richardson; Jerry Scutts; Brian Pymm; Maj-Gen. Carl G. Schneider, USAF Retd.; Geoff Stuart; Wg Cdr C. C. 'Jeff' Jefford, MBE, BA for kind permission to quote Dickie Dickinson's story as related in the *RAF Historical Society Journal No. 21*, 2000.

This book is dedicated to the memory of Lt-Col Reg W. Adams and all other Sabre pilots who have made their last flight.

Text written by Martin W. Bowman
Profile illustrations created by Dave Windle
Cover painting by Jim Brown – The Art of Aviation Co. Ltd

First published in the UK in 2004
by Airlife Publishing, an imprint of The Crowood Press Ltd

British Library Cataloguing-in-Publication Data
A catalogue record for this book
is available from the British Library

ISBN 1 84037 411 X

Printed in Malaysia

Contact us for a free catalogue that describes the complete range of Airlife books for pilots and aviation enthusiasts

Airlife

An imprint of The Crowood Press Ltd
Ramsbury, Marlborough, Wiltshire SN8 2HR
E-mail: enquiries@crowood.com

www.crowood.com

Contents

F-86/FJ Timeline

18 May 1945
North American Aviation (NAA) receives a contract for three NA-140 prototypes under the USAAF designation XP-86. The same day, 100 NA-141s (subsequently reduced to 30 aircraft) are ordered for the US Navy as the FJ-1

1 November 1945
Redesigned XP-86 airframe, featuring sweepback on all flying surfaces, is accepted by the USAAF

8 August 1947
First of two flying XP-86 prototypes completed

1 October 1947
XP-86 45-59597, flown for the first time by North American test pilot George S. Welch

28 May 1948
First three production Sabres delivered to USAF

15 September 1948
At Muroc, Maj. Johnson, flying F-86A-1 47-611, successfully raises the World Air Speed Record to 1079.61 km/h (670.84 mph)

January 1949
The first operational F-86As are delivered to the 94th FS, 1st FG at March AFB, California

4 March 1949
F-86 officially named Sabre

early 1949
Canada decides on the Sabre to form the nucleus of a new RCAF jet-fighter force. Canadair is chosen to build the aircraft under licence

December 1950
F-86A replaced in production by the F-86E-1-NA

15 December 1950
4th FIW begins operations over North Korea

17 December 1950
Lt-Col Bruce H. Hinton, CO, 336th FIS, 4th FIW shoots down the first MiG to be destroyed by a Sabre in the Korean War

20 May 1951
Capt. James J. Jabara becomes the first jet ace of the Korean War

October 1951
Australian government obtains a manufacturing licence for the Sabre

Spring 1952
RAF to receive 370 (later upped to 430) Canadian-built Sabres under the Mutual Defense Assistance Pact (MDAP)

June 1952
39th FIS, 18th FIW, equipped with F-86F Sabres, arrives in theatre

18 November 1952
Capt. J. Slade Nash shatters the World Speed Record with a speed of 1124.104 km/h (698.505 mph) in F-86D-20 51-2945, over a 3-km (1.86-mile) course at Salton Sea, CA

18 May 1953
F-86A (Sabre 1) 19101 flown around the 100-km (62.14-mile) course at Edwards AFB by Jacqueline Cochran, to set a new Women's World Air Speed Record of 1050.152 km/h (652.552 mph). The same day she becomes the first woman to exceed Mach 1. On 23 May, with wingtip tanks added, she flies over the 500-km (310.69-mile) course at 949.926 km/h (590.273 mph) and on 3 June flies at 1078 km/h (670 mph) over the 15-km (9.32-mile) straight course

16 July 1953
Lt Col William Barnes, AMC production test pilot, sets a new 1151.771-km/h (715.697-mph) record in F-86D-35 51-6145 at Salton Sea, CA

September 1954
Maj. John L. Armstrong, in an F-86H, sets a new World Speed Record of 1044 km/h (649 mph) for the 500-km (310.69-mile) closed circuit. Another F-86H, flown by Capt. Eugene P. Sonnenberg, sets a new 1114.9 km/h (692.8-mph) record for the 100-km (62.14-mile) closed circuit

16 March 1956
Last F-86H-10 (53-1528) delivered to the USAF

30 August 1956
Two RCAF Sabre 6s, piloted by Flt Lt R. H. Annis and F/O R. J. Childerhouse, fly from Vancouver to Dartmouth, Nova Scotia, a distance of 4409 km (2,740 miles) in 5 hours 30 seconds, including a 10 minute stop, to set a new record

28 October 1956
F-86F reinstated in production as the NA-227 or F-86F-40

1 October 1956
American-supplied F-86Fs used to form the JASDF's first tactical jet fighter unit at Hammatsu

28 December 1956
F-86F-40 55-5047, the last of 5,035 Inglewood, CA-built Sabres accepted. Including 1,175 built at Columbus, 6,210 Sabres had been built in the US

1. Prototypes and Development

When in early November 1950 5th Air Force Lockheed F-80 Shooting Stars first encountered the Mikoyan-Gurevich MiG-15 'Fagot' over northwest Korea, it was the first time since early World War Two that US forces lacked air superiority in a theatre of war. The appearance of the Soviet-built, swept-wing jet fighter sent shockwaves through the US military. It was 121 km/h (75 mph) faster than any aircraft in the US Air Force (USAF) inventory (the Soviet design also benefited from Britain's world lead in jet propulsion development when, in 1946, 25 Rolls-Royce Nene turbojets were sold to the Soviets as part of the Anglo-Soviet Trade Agreement). A month later the USAF had regained the initiative thanks largely to the F-86A Sabre, which generally proved to be the equal of the MiG-15, although the lighter enemy jet proved to have the edge over the F-86A in climb and altitude performance and was more manoeuvrable.

The Sabre's development had started in mid-1945 at North American Aviation (NAA) at Inglewood, California, where a team headed by

The XJF-1 is shown here at Inglewood, California, on 21 August 1946. The straight-winged XFJ-1 flew for the first time on 27 November 1946, powered by a J35-GE-2 turbojet. *(via Robert Jackson)*

The F-86A production line at Inglewood, California. *(NAA via Larry Davis)*

Ray Rice and Edgar Schmued began work on two company projects, the NA-134 carrierborne jet fighter and the NA-140 land-based version for the US Army Air Force (USAAF). On 18 May 1945, NAA received a contract to build three NA-140 prototypes under the USAAF designation XP-86. It also received a contract to build 100 NA-141 (NA-134) jet fighters for the US Navy (USN) as FJ-l Fury aircraft. The XFJ-1 flew for the first time on 27 November 1946, powered by a 16.99-kN (3,820-lb st) J35-GE-2

turbojet. Maximum speed was 858 km/h (533 mph) at sea level and 872 km/h (542 mph) at 4877 m (16,000 ft) and it was capable of attaining Mach 0.87 in a dive. Though performance was inferior to that of the XP-86 in all but range, the FJ-1 was nonetheless the first jet fighter to enter USN service and it became the first American jet to operate from a carrier in squadron strength. However, only one USN fighter squadron ever operated the FJ-1, which was a conversion of an existing air force design. Its nearest competitor, the McDonnell FH-1 Phantom, was the first American jet aircraft to be designed for aircraft-carrier operation from the

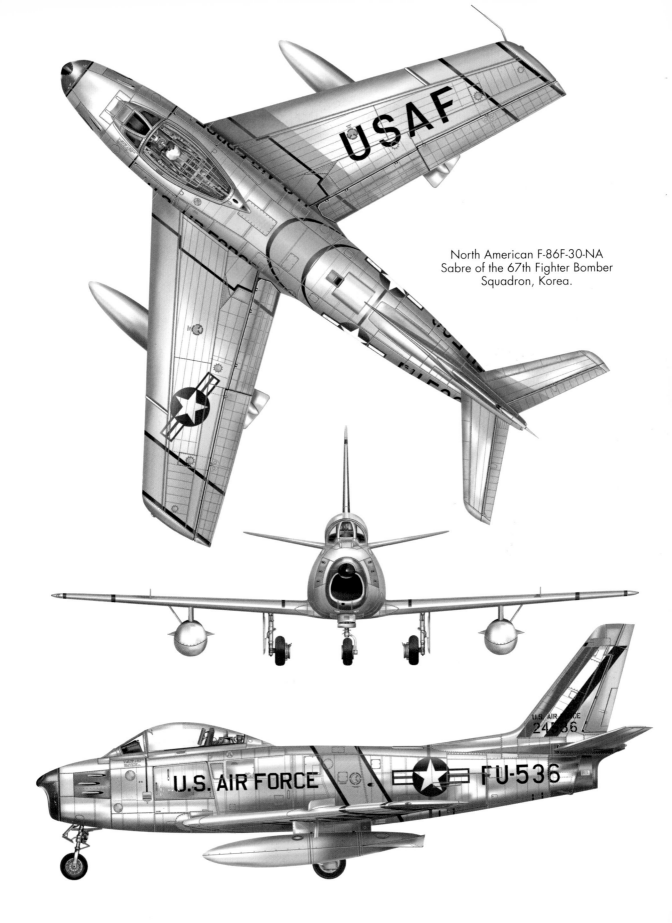

North American F-86F-30-NA
Sabre of the 67th Fighter Bomber
Squadron, Korea.

outset. Also, the FJ-1's high fuel consumption required much larger fuel storage aboard carriers. The 1945 order for 100 production FJ-ls was subsequently reduced to a requirement for just 30 aircraft. Some 24 of these were operated by VF-5A (later VF-51), which received the first example at NAS North Island, San Diego, California on 15 November 1947. Production FJ-1s differed from the three XFJ-1 prototypes in having an Allison-built J35-A-4 turbojet rated at 17.79-kN (4,000-lb st). Gross wing area was also increased by the provision of wing-root leading-edge extensions. VF-51 put the FJ-1 through its carrier compatibility trials at North Island and aboard the USS *Boxer* in 1948, while sister squadron VF-17 evaluated the FH-1 Phantom. On 16 March 1948, Cdr Evan 'Pete' Aurand, Commanding Officer (CO) of VF-51, had the distinction of making the first carrier landing in an FJ-1 when he landed aboard *Boxer*. Further carrier trials aboard *Princeton* revealed a weakness in the Fury's undercarriage when landing aboard ship and overall the aircraft showed scant suitability for carrier operations.

The prototype XFD-1 (ordered as the FD-1, which was later redesignated as the FH-1) flew for the first time on 25 January 1945, powered by two Westinghouse J30 turbojets. On 21 July 1946 an FD-1 carried out the first US jet aircraft carrier trials, and as a result of these a production order for 100 aircraft, later reduced to 60, was placed with McDonnell. On 15 May 1948, VF-17A with 16 FH-1 Phantoms, became the first fully operational USN jet fighter squadron at sea, forming part of the Air Wing aboard USS *Saipan*. The McDonnell F2H Banshee then entered service with the USN in 1949, and in May 1951 the FJ-1s of VF-5A were withdrawn from first-line service and assigned to Reserve units. That same month the USN received its first examples of the much more advanced Grumman F9F-2 Panther jet fighter-bomber when they began equipping VF-51.

XP-86 for the USAF

Meanwhile, a mock-up of the XP-86 was built and in June 1945 the USAF approved it. NAA had estimated that the XP-86 would have a

XFJ-1 (NA-134) Bureau of Aeronautics Number (BuNo.) 39053, one of three prototypes built, is the central machine in this image. The 1945 order for 100 production FJ-1 Furys was reduced to a purchase of just 30 aircraft and 24 of these were operated by VF-5A (later VF-51), which received the first example at NAS North Island, San Diego, California, on 15 November 1947. *(via Larry Davis)*

This VF-5A FJ-1 Fury suffered a nosewheel collapse during landing aboard its carrier. *(via Larry Davis)*

maximum sea level speed of 924 km/h (574 mph) at a gross weight of 5216 kg (11,500 lb) but the USAF specification called for a maximum speed of 600 mph (966 km/h). At Republic Aviation Corporation, the XP-84 jet fighter-bomber prototype had an estimated performance that greatly exceeded that of the XP-86, so NAA had to somehow squeeze more performance out of its design and quickly. It was at this juncture that material on German research into swept-wing design became available for study. The swept-wing concept was not new. It had been identified by German aerodynamicists in the

mid-1930s, but the early American and British jet aircraft were conventional, straight-wing designs. US and British designers only really became aware of the benefits of swept-wing design when captured German research data became available and a complete Messerschmitt Me 262 wing assembly was sent to Los Angeles. The Me 262 had a low-mounted wing of moderate sweepback (18° 32' on the leading edge) to delay the drag rise caused by the

XFJ-2B (NA-185) Fury BuNo. 133756, was used purely as an armament test airframe at the Ordnance Test Station, Inyokern. It tested a four Colt Mk 12 20-mm cannon installation and was not navalised. *(via Larry Davis)*

formation of shock waves, and the wing assembly enabled NAA to assess the effect of leading-edge slots on low-speed stability. The NAA design team led by Project Aerodynamicist L. P. Greene carried out many hundreds of wind tunnel tests with model wings at varying angles of sweep and concluded that the XP-86's limiting Mach number could be raised to 0.875 by incorporating wing swept at an optimum angle of 35°. The USAAF accepted the redesigned

XP-86 airframe on 1 November 1945 and the swept-wing proposal received its final approval on 28 February 1946. The redesigned XP-86 featured sweepback on all its flying surfaces – both the fin and tailplane having a 35° sweepback – and the tailplane was set at a 10° dihedral angle. The fuselage was of conventional stressed-skin construction and was built in two main sections to aid installation, inspection and removal of the turbojet. The rear fuselage was of simple construction and the rearmost bay was made of stainless steel. Speed brakes, operated by electrically controlled hydraulic jacks and served by an engine-driven pump, were fitted on each side of the rear fuselage. The large clearvision cockpit canopy used on the XP-86 was retained and a loop aerial and cockpit pressure regulator were fitted to it. The aircraft's ejection seat was an NAA development of a government-sponsored design.

USAF contract

In December 1946 the USAF placed a contract for 33 P-86A-1-NA aircraft. On 8 August 1947 45-59597, the first of two flying XP-86 prototypes (the third being a static test machine) was completed and transferred to Muroc (now Edwards Air Force Base (AFB)) for several weeks of assembly, systems testing and taxiing trials. A 17.79-kN (4,000-lb st) Chevrolet-built J35-C-3 turbojet powered the aircraft. Finally, on 1 October, 45-59597 was flown for the first time by NAA test pilot George S. Welch. The flight lasted 50 minutes, but the nosewheel refused to lock in the down position and Welch only finally succeeded in getting it down with little time to spare before landing safely. Welch completed the

Phase I testing of the XP-86 in just 30 flying hours, before 45-59597 was presented to the USAF for Phase II testing. This was undertaken by Maj. Kenneth O. Chilstrom, who made eleven flights totalling 10 hours 17 minutes on 2–8 December 1947. Test flights were made at an all-up weight of 6255 kg (13,790 lb). The XP-86 was now powered by an Allison-built J35-A-5 developing 17.43 kN (3,920-lb st), which gave the aircraft a maximum speed of 964 km/h (599 mph) at sea level, 995 km/h (618 mph) at 4267 m (14,000 ft) and 925 km/h (575 mph) at 10668 m (35,000 ft). Testing revealed that on take-off the XP-86 tended to sink, so the take-off angle-of-attack was increased to 13° and maintained until safe flying speed was reached. Unstick speed was about 201 km/h (125 mph) after a roll of 920 m (3,020 ft) in still-air conditions. Climb at sea level was 1219 m (4,000 ft) per minute. The XP-86 could reach 6096 m (20,000 ft) in just over 6 minutes and 9144 m (30,000 ft) in just over 12 minutes. There

F-86A-5 49-1241, was one of a batch of 333 (NA-161) Sabres (49-1007/1339) from the second contract. NA-161s were delivered between October 1949 and December 1950. The A-5 was powered by an uprated J47-GE-3 engine of 23.12 kN (5,200 lb st). *(via Robert Jackson)*

These F-86As of the 93rd FIS, 81st FIW, Western Air Defense Force, Kirtland AFB, New Mexico, were diving for a practice firing mission against an aerial target during the First Annual Air Defense Command (ADC) Gunnery Meet, held between 5 and 11 March 1950. (NAA via Jerry Scutts)

18 May 1948. Power was provided by the more powerful J47-GE-1 turbojet developing 21.57 kN (4,850 lb st). The third XP-86 (45-59599) also flew in that May and was accepted by the USAF on 17 December. Serial number 47-605 was the first of the F-86A-1 production aircraft and 47-606 and 47-607 that followed were delivered to the USAF on 28 May 1948. These were the first F-86s armed with six 0.50-in (12.7-mm) M-3 machine-guns. Flush-fitted automatic gun-muzzle doors which opened and closed when the machine-guns were fired were later deleted and replaced by plastic plugs, which were blasted clear by the first fired rounds. The pilot fired the guns using a Mk 18 lead computing sight that included a gyro and fixed sighting system. Targets were identified and the span selector lever set to correspond with the target aircraft's span. When the target appeared within a circle of six diamond images on the reflector the range control was rotated until the circle diameter was identical in size to the target. In one second the gunsight automatically computed the required lead and the guns were fired. The more complex A-1CM gunsight was fitted in place of the Mk 18 sight on the last 24 F-86As and on earlier F-86A-5s modified to the A-6 and A-7 standard. The A-1CM was coupled with an AN/APG-30 radar installed in the upper lip of the nose intake. The radar had a sweep range of about 137–2743 m (150–3,000 yards) and automatically locked on to targets and tracked them. Below 1829 m (6,000 ft) altitude, ground clutter reduced the effectiveness of the radar ranging and manual ranging with a sight range dial was used.

Larger orders

On 10 June 1948 the USAF ordered a third batch of 333 aircraft designated F-86A-10, -15 and -20,

was no doubt that the XP-86 was far superior to any other jet fighter design. On 26 April 1948, George Welch exceeded the speed of sound in a shallow dive. The XP-86 thus became the first combat aircraft in the Western world capable of supersonic performance and the second American aircraft to exceed Mach 1, for on 14 October 1947 Capt. Charles 'Chuck' Yeager had reached sonic speeds in the Bell X-1 rocket-powered aircraft.

The second prototype (45-59598), designated XF-86A by the USAF, flew for the first time on powered by either the J47-GE-7 or -13 turbojet. To meet a USAF need for larger tyres, NAA decided to increase the F-86's fuselage width by 17.78 cm (7 in) to accommodate them when retracted. The changes led to a re-designation to F-86B but although 188 were ordered, the F-86B never flew because higher-pressure tyres and new brake designs became available. The F-86B contract was transferred to a new batch of F-86As, designated F-86A-5-NA aircraft, to be fitted with the improved standard-size tyres. Deliveries of the A-5 model began in late March 1949 when the USAF accepted 48-129. The A-5 was powered by an uprated J47-GE-3 engine of

23.12 kN (5,200 lb st). Outwardly the aircraft differed from the F-86A in having a V-shaped, instead of rounded, windscreen.

In 1946 meanwhile, Strategic Air Command (SAC) originated a requirement for a 'penetration fighter' to escort its huge Convair B-36 Peacemaker bombers operating from bases in western Germany to potential targets in the Soviet Union. Equally, the new fighter had to have the ability to range ahead of the bombers and establish air superiority en route to the target areas. Lockheed, McDonnell and NAA all put forward proposals. Although the Lockheed XF-90 was aerodynamically advanced, heavily

F-86A-5s (NA-161s) 49-1197 and 49-1289 of the 116th FIS, 81st FIG after arriving in England on 27 August 1951. The 116th FIS's aircraft became the first Sabres to cross the Atlantic and the squadron itself became the first USAF unit to be stationed in Britain since World War Two. It was based at Shepherd's Grove, Suffolk. *(via GMS)*

armed and had the required range, it was seriously underpowered, while the McDonnell XF-88 had an inferior combat radius and operational ceiling. NAA's proposal, originally designated F-86C, potentially offered the best solution and on 17 December 1947 the F-86C (NA-157) long-range fighter was ordered. It retained the basic Sabre wing, although the span was increased to 11.81 m (38 ft 9 in) and the axial-flow J47 engine was replaced by a 27.79-kN (6,250-lb) thrust Pratt & Whitney XJ48-P-1 turbojet which delivered 38.91-kN (8,750-lb) thrust with afterburner. To accommodate the new larger-diameter, centrifugal-flow engine the fuselage was enlarged in cross-section and re-contoured. To achieve the maximum possible reduction in drag the fuselage was stream-contoured, and twin-wheel main undercarriage units were fitted to support the new fighter's 11567-kg (25,500-lb) loaded weight. Fuel capacity was increased to 5985 litres (1,581 US gal). Twin NACA (National Advisory Commitee for Aeronautics) flush-type intakes, which permitted the removal of the nose intake in favour of an AI (air-intercept) radar installation, were fitted on the sides of the cockpit. Another distinguishing feature was twin wheels on the main landing gear. Only two aircraft were built to the NA-157 specification, and were much larger than the standard F-86 and were re-designated YF-93A. Some 118 F-93A aircraft were ordered in June 1948, but in January 1949 this contract was cancelled to make fiscal funds available for more B-36 bomber aircraft. George Welch flew the first YF-93A (48-317) on 25 January 1950. This and the second YF-93A prototype (48-318), which flew in May 1950, were eventually presented to NACA's Ames Laboratory at Moffett Field for comparison tests of the flush air intakes with the scoop intakes also trialled on the prototypes.

Sabre into service

Early in 1949 deliveries of the first F-86As were made to the 94th Fighter Squadron (FS), 1st Fighter Group (FG) March AFB, California, where they replaced F-80s. The 1st FG was assigned to SAC, its role being the defence of SAC air bases. The F-86A was an aircraft with no name so the 1st FG sponsored a competition to find a suitable appellation; 78 names were submitted and on 4 March 1949 the F-86 was officially named Sabre. By the end of May, the 1st FG had received 83 F-86As and the 27th and 71st FSs were also equipped with the new model. On 16 April 1950 the 1st FG was re-designated the 1st Fighter-Interceptor Group (FIG) and on 1 July was transferred to Continental Air Command. Its primary task was to protect the aircraft production industries in the Los Angeles area. Continental air defence was of primary importance as the Cold War between East and West intensified. The Tupolev Tu-4 'Bull' (a Soviet-built B-29 derivative) long-range bomber had entered service with the Dalnaya Aviatsiya (Long-Range Aviation, DA) in May 1949 and that same year the Soviets had detonated their first atomic device. Attacks on American industrial centres by nuclear-armed Tu-4s now became an alarming possibility.

On 1 December 1951 the 1st FIG moved to Norton AFB, California, where it was de-activated on 6 February 1952. Deliveries of F-86A-10, -15 and -20 (49-1007/49-1339) aircraft were made during October 1949–December 1950 and equipped four more Fighter Interceptor Groups – the 4th FIG at Andrews AFB, Maryland; the 33rd FIG at Otis AFB, Massachusetts; the 56th FIG at Selfridge AFB, Michigan, and the 81st FIG at Kirtland AFB near Albuquerque, New Mexico. The 4th FIG was responsible for the air defence of Washington, DC, while the 81st FIG protected the nuclear weapons production facility at Los Alamos. During August–September 1951, the 81st FIG deployed to Great Britain, the first squadron to arrive being the 116th Fighter-Interceptor Squadron (FIS), which landed at Shepherd's Grove, Suffolk on 27 August. It was joined there a short time later by the 92nd FIS, and the 91st FIS arrived at Bentwaters, Suffolk on 3 September.

2. Operational History:
Countering the MiG-15 over Korea

In SE Asia the Cold War became hot when on 25 June 1950 Communist North Korean forces invaded the Republic of South Korea. United Nations (UN) land-based counteroffensives soon reached the banks of the Yalu River, which formed the border with the People's Republic of China. Unfortunately, during the first week of November, China became embroiled in the conflict, sending masses of ground troops across the bridges on the Yalu. That month the overwhelming balance of air power in North

On 1 and 9 November 1950, 75 F-86Es were shipped in batches from Alameda, California, on the escort carriers *Cape Esperance* (pictured) and *Sitkoh Bay* to Korea, to re-equip two F-80C squadrons of the 51st FIW. *(John Henderson via Larry Davis)*

F-86A-5 49-1327, of the 4th FIW, is shown undergoing an engine change at K-13 Suwon in April 1951. The area under the B-29 wing was used as the engine shop! *(via Larry Davis)*

Korean airspace changed dramatically with the intervention by Soviet-built MiG-15 jets of the Chinese Communist Air Force. The UN had no combat aircraft in the theatre capable of meeting the MiG-15 on equal terms. On 26 November, 18 Chinese divisions entered the battle and soon the UN Forces were in headlong retreat.

On 8 November Gen. Hoyt S. Vandenberg, USAF Chief of Staff, had offered to release the 4th Fighter-Interceptor Wing (FIW) and its

F-86A-5s (49-1129 *Minimum Effort*, right) of the 4th FIW take off from K-14 Kimpo on 17 December 1950, the day the first MiG kill in Korea was recorded. *(via Larry Davis)*

F-86A-5 Sabres and the 27th Fighter-Escort Wing (FEW) and its Republic F-84E Thunderjets, for operations in Korea. Gen. Earle E. Partridge, commanding US Far East Air Forces (FEAF) and Gen. George E. Stratemeyer, commanding 5th Air Force, immediately accepted the offer. The 4th FIW, commanded by Col John C. 'Whips' Meyer, a World War Two fighter pilot with 24 victories, was at New Castle County Airport, Wilmington, Delaware, where the 4th was assigned to the Eastern Air Defense Force. On 11 November the Sabres were flown to San Diego, California, where, on 29 November, the aircraft of the 334th and 335th FISs were deck-loaded aboard the escort carrier *Cape Esperance* and the 336th FIS on board a fast tanker for Yokosuka, Japan. Advance personnel were sent to Japan by air, and the main contingents followed by rail and then by naval transport. The Sabres arrived on 13 December and an advance detachment flew to Kimpo (K-14) a few miles south of the 38th Parallel and to the north of Seoul, the South Korean capital. They were soon in action. Four F-86A-5s of the 336th FIS engaged

Captain James J. Jabara of the 334th FIS, 4th FIW, is given a hero's welcome on 20 May 1951. He had just shot down his fifth MiG in F-86A 49-1319, to become the first jet ace of the Korean War. *(via Larry Davis)*

in combat with the MiG-15 for the first time on 17 December 1950. Lt-Col Bruce H. Hinton, CO, 336th FIS, shot down the first MiG and the other three, too fast for the pursuing F-86s, made for the border. Hinton's was the first of 792 MiGs to be destroyed in the Korean War.

More Sabre-MiG combat

Sabres and MiGs clashed again on the morning of 19 December, when Lt-Col Glenn T. 'Eagle' Eagleston damaged one of the enemy jets. He also claimed a probable two days later when eight Sabres tried to intercept two MiGs flying at 10363 m (34,000 ft). On the morning of 22 December, the first Sabre was shot down in combat with a MiG, but the F-86 pilots gained revenge later that day when the 4th FIW destroyed six MiGs in one combat. On New Year's Day 1951 the Chinese launched a new invasion of South Korea, which succeeded in removing the UN Forces from Seoul. Next day, the 4th FIW had to evacuate to Johnson AFB in Japan. In desperation, a detachment of Sabres arrived at Taegu on 14 January to begin ground attack sorties over the enemy lines. However, they found only limited success – just two 5-in (127-mm) rockets could be carried in addition to the Sabre's machine-guns because of the need to carry drop tanks – and the extreme operating range meant that pilots only had a short time to locate and hit targets. By 31 January, though, the F-86s had completed 150 valuable close-support sorties.

UN Forces launched a counteroffensive which recaptured Seoul and the airbases at Suwon and Kimpo, but they were so badly damaged that the USAF had to wait until engineering battalions made them habitable again. On 22 February the 334th FIS, 4th FIW flew into Taegu (K-5) in the south of the country, but it was too far away for

F-86A-5 48-215 suffered a nosewheel collapse at Johnson AB, Japan, in the summer of 1951. *(via Larry Davis)*

the Sabres to hunt MiG-15s over MiG Alley. The furthest the Sabres could fly was to Pyongyang and the area which became known as MiG Alley north of a line between Wonsan and Pyongyang. Finally, on 10 March, the single concrete runway at Suwon (K-13) was able to take jets again and the 334th FIS occupied the battered airfield. Meanwhile, the 336th FIS, which flew in from Japan, occupied Taegu.

Despite their experience, Sabre pilots found it increasingly difficult to maintain air superiority over the Communist air force. Enemy tactics in April showed a marked reluctance to continue with large formations of 50-plus MiGs. These were replaced by more flexible formations of 16 fighters in four flights of four, so the USAF dispatched small patrols of 12 F-86As backed up by supporting flights, which could be called upon when the first group had flushed the MiGs. On 22 April four enemy jets were shot down and four more damaged. One of the victories went to Capt. James 'Jabby' Jabara of the 4th FIW, who shot down his fourth MiG of the war. Jabara, who had downed 3.5 German aircraft in World

War Two, became the first jet ace of the Korean War on 20 May when 50 MiGs intercepted a patrol of 14 F-86As over Sinuiju. Jabara, who was in a second wave of 14 Sabres, tacked on to three MiGs at 10668 m (35,000 ft) and singled out the last one. He blasted it with cannon fire and MiG the pilot bailed out at 3048 m (10,000ft), just before the fighter disintegrated. Jabara climbed back to 6096 m (20,000 ft) and bounced six more MiGs. He fired at one and the enemy fighter began to smoke and catch fire, before falling into an uncontrollable spin.

Mounting MiG strength

By the summer of 1951 it was estimated that the Chinese People's Air Force (CPAF) had 1,050 Soviet-built combat aircraft, including 445 MiG-15s (when only 89 F-86A Sabres were available and just 44 of them in Korea) and new

F-86A-5 49-1334 of the 335th FIS, 4th FIW, was hit by a 'Bedcheck Charlie' night raider on the ramp at K-13 Suwon on 17 June 1951. These 'Bedcheck Charlie' raids were usually flown by old, slow aircraft, typically the Polikarpov Po-2 'Mule' biplane. They proved extremely difficult to combat. *(Leo Fournier via Larry Davis)*

airfields were under construction in the Antung area to accommodate 300 of the jet fighters. The enemy pilots sometimes now flew in singles and pairs, with drop tanks fitted to their MiGs, as far south as the 38th Parallel. Their attacks bore all the hallmarks of being carried out either by Red Chinese instructors or Soviet pilots – 'Honchos' (from the Japanese word meaning 'boss') as the Americans called them. A worried Gen. Otto P. Weyland, who assumed command of FEAF on 10 June, saw the immediate need for four more Sabre Wings. Since June/July, FEAF's F-86As

In an attempt to hide them from 'Bedcheck Charlie', these F-86As of the 4th FIG at K-14 Kimpo were placed under camouflage netting in the summer of 1951. *(via Larry Davis)*

had begun to be replaced by F-86Es on a one-for-one exchange basis, but this was a very slow process. (Beginning in September, a few F-86Fs filtered through to the 4th FIW but the F-86A was not completely replaced in the 4th FIW inventory until July 1952). Weyland wanted two F-86E wings to be sent immediately to Korea,

These F-86 Sabres of the 4th FIW were photographed at K-13 Suwon in the summer of 1951. *(via Larry Davis)*

and two to Japan, where they would help deter possible Chinese attacks on Japanese bases should the Soviets supply the CPAF with Ilyushin Il-28 'Beagle' jet bombers. Weyland's requests for F-86Es, however, fell on deaf ears in Washington, where the CPAF build-up was viewed as being purely defensive. Washington therefore argued that any major increase in UN air reinforcements would be seen as a prelude to all-out air warfare against China. In any case, the USAF did not want to weaken an already below strength Air Defense Command. Weyland received one F-84 Wing, no other Sabre wings were forthcoming. By September when a new Communist air offensive began, there were no less than 525 MiG-15s in the enemy inventory.

Nevertheless, the Sabres continued to knock down MiGs at a good rate. That month the 4th FIW sighted 1,177 MiG sorties over North Korea and engaged 911 MiGs in combat. Even though the odds were usually stacked against them the Sabres scored some impressive victories. On 2 September, 22 F-86s battled with 44 MiGs between Sinuiju and Pyongyang and shot down four. Seven days later, in a battle between 28 Sabres and 70 MiG-15s, Capts Richard S. Becker and Ralph D. 'Hoot' Gibson claimed a MiG each to become the second and third aces of the Korean War. Three F-86s were lost in these two air battles. Despite the successes, by the end of September the Communist air force posed a very real threat to American fighter-bomber operations in the North and Weyland once again requested an additional Sabre wing, or at least the conversion of one of the existing F-80 wings to F-86Es. His request was rebuffed and, as a result, Gen. Frank F. Everest, commander 5th AF since 1 June, had no option but to withdraw his fighter-bombers from MiG Alley and order them instead to concentrate on a zone between Pyongyang and the Chongchon River.

In October patrols over MiG Alley were stepped up and in the first two weeks of operations the 4th FIW claimed 19 MiGs (including nine on 16 October) destroyed. During October 1951 no less than 2,573 MiG sorties were sighted and of these 2,166 were intercepted and 32 MiGs claimed destroyed, although the USAF lost ten jet fighters and five Boeing B-29 Superfortresses. Flushed with success, the Communists relocated their fighters further south of the Yalu and based fleets of bombers at Sinuiju. The Communists also boasted a new fighter, the MiG-15bis, which had an uprated VK-1 turbojet in place of the original RD-45. In good hands the new jet easily

F-86A 49-1080 was flown by Lt-Col Glenn T. 'Eagle' Eagleston, CO, 334th FIS, 4th FIW. Note the 5-in (127-mm) high-velocity aircraft rocket (HVAR) just inboard of the drop tank. *(via Larry Davis)*

F-86E-10 51-2800 *Liza Gal/El Diablo*, which was assigned to Capt. Chuck Owens of the 336th FIS, 4th FIW, at K-14 in 1952, shows eight MiG kill markings and records 15 trucks and a tank destroyed.

F-86E-10 51-2801 *Jimmie Boy II* and another 25th FIS Sabre sit on five-minute alert at K-13 Suwon early in June 1952. *(both via Larry Davis)*

outclassed the F-86A. The 4th FIW, including the 335th FIS in Japan, moved up to Kimpo but could not stave off the threat posed by the MiGs on its own. Unless more F-86Es arrived soon then the UN would lose air superiority in Korea.

Shipping Sabres to Korea

Gen. Hoyt S. Vandenberg, USAF Chief of Staff, arrived in the Far East late in October. The Chinese air force had, he said, 'almost overnight become one of the major air powers of the world'. Vandenberg ordered the transfer from Air Defense Command of 75 F-86Es, together with air and ground crews and full supporting equipment, to the Korean theatre to re-equip two squadrons of F-80Cs of the 51st FIW. On 1 and 9 November the 75 F-86Es were shipped from Alameda, California on the escort carriers *Cape Esperance* and *Sitkoh Bay* to Korea. The 51st FIW re-formed at Suwon on 6 November and an equal number of F-80C crews returned Stateside

in exchange. That same day legendary World War Two fighter ace Col Francis S. 'Gabby' Gabreski, who had 28 confirmed German aircraft kills and had added three MiGs to his score while flying missions as deputy CO of the 4th FIW in Korea, assumed command of the new Sabre wing. (Gabreski went on to add 3.5 victories while commanding the 51st. The half MiG was shared with Maj. Bill Whisner, another World War Two ace, who finished the Korean War with 5.5 kills).

In November 1951 the 4th FIW, now commanded by Col Harrison R. Thyng, another World War Two ace with 11 enemy aircraft to his credit, claimed 14 MiGs destroyed. On 18 November four Sabre pilots sighted 12 MiG-15s in dispersals on Uiju airfield, and while two of them provided top cover the other two, Capt. Kenneth D. Chandler and Lt Dayton W. Ragland, made a low-level strafing attack that destroyed four MiGs and damaged several

F-86A-5 49-1272 *Wham Bam*, flown by Lt Martin Bambrick, and F-86E-10 51-2769 *Bernie's Bo* (with two kill markings), of the 335th FIS, 4th FIW, were photographed at K-14 Kimpo. Bambrick destroyed a MiG while flying *Wham Bam* on 4 September 1952. *Bernie's Bo* was flown by Capt. Robert J. Love (later the 11th jet ace of the war with 6 kills) and subsequently by Capt. Clifford D. Jolley (7 kills), who had to eject from the aircraft on 4 July 1952. *(via Robert Jackson)*

These F-86Es hailed from the 335th FIS, 4th FIW. Note the Indian Head insignia and the black-edged yellow recognition fuselage band, which appeared late in 1951. An additional black-edged yellow band across the fin acted as a Wing marking. The 335th FIS was the top scoring unit of the Korean War, with 14 of the USAF's 39 Sabre aces and its pilots claiming no less than 218.5 air-to-air victories. 50-625, 'Bones' Marshall's F-86E, has 18 stars to denote MiG kills scored by various pilots in this Sabre. *(Larry Davis)*

others. Four MiGs were shot down in a major air action on 27 November, one of them by Maj. Richard D. Creighton, who became the fourth jet ace of the Korean War. In the afternoon of 30 November the biggest air combat success so far took place, when 31 Sabres of the 4th FIW, led by Col Benjamin S. Preston, sighted a formation of 12 Tupolev Tu-2 'Bat' bombers escorted by 16 piston-engined Lavochkin La-9 'Fritz' fighters and 16 MiG-15s heading for the island of Taehwado in the Yellow Sea, where Republic of Korea (RoK) forces were fighting North Korean marines. Maj. George A. 'Curly' Davis, Jr, CO 334th FIS, and Maj. Winton W. 'Bones' Marshall, CO, 335th FIS, became jet aces that day. Davis destroyed three of the bombers, damaged another, and downed a MiG. Marshall shot down one bomber, damaged another, and downed an La-9. All told, the Sabres destroyed eight Tu-2s, three La-9s and a MiG-15.

New Year air offensive

The 51st FIW's F-86E Sabres went into action on 1 December. On 2 and 4 December Sabre pilots claimed ten MiGs, two of them being claimed by

51st FIW pilots. Maj. George Davis of the 4th FIW destroyed two more MiGs on 5 December. On the 13th he claimed another four MiGs – a Korean record – when 4th FIW Sabres took on 145 MiG-15s in air battles along the Yalu and destroyed 13 of them in a bitter engagement. Although MiG-15s continued to appear over North Korea in large numbers, they avoided combat below 9144 m (30,000 ft). Just three more MiGs were destroyed by the end of December. One was claimed by the 4th FIW on 14 December and the other two by the 51st FIW on 15 and 28 December. A total of 28 MiGs was shot down in December for the loss of seven Sabres.

In January 1952, the Communists launched a New Year air offensive, sending as many as 200 MiGs across the Yalu simultaneously at speeds up to Mach 0.99. Just five F-86s were lost in the air, but during January–February 4th FIW F-86A Sabres downed only 11 MiGs, the Communists having raised their operational altitude to avoid battle with the F-86As which still mainly equipped the 4th FIW. The F-86Es of the 51st FIW meanwhile, could climb to 13716 m (45,000 ft) before engaging the enemy. They took

RF-86F 52-4808, of the 15th Tactical Reconnaissance Squadron (TRS), 67th Tactical Reconnaissance Wing (TRW), is shown with black-painted fake gun ports to scare off MiGs. The cheek fairing over the compartment below the cockpit contained a pair of 20-in (508-mm) K-24 cameras mounted lengthwise with a mirror arrangement to provide vertical coverage. 52-4808 was scrapped at Kisaruzu in December 1957. *(USAF)*

F-86E 50-0653, of the 25th FIS, 51st FIW, is shown in Korea in 1953. This Sabre (as N5637V) is now displayed at the USAF Museum, Hickam AFB, HI. *(via Larry Davis)*

F-86F Sabre 602/J of No. 2 Squadron, SAAF, part of the USAF's 18th FBW, is seen over Korea with Gloster Meteor Mk 8 A77-871 of No. 77 Squadron, Royal Australian Air Force (RAAF). 602 was one of the first five F-86Fs (renumbered 601 to 605) delivered to the SAAF squadron in late January 1953, to replace its F-51Ds. In all, No. 2 Squadron received 18 F-86Fs and resumed operations on 16 March 1953. *(Grp Capt. Dickinson)*

on the MiGs on equal terms and shot down 36 aircraft during the same period for the loss of only two F-86s. One of these Sabre losses occurred on 10 February. Maj. George A. Davis, Jr, of the 4th FIW, who was the leading jet ace with 12 kills, was leading 18 Sabres on an escort mission to the railway yards at Kumu-ri, when he spotted a formation of MiGs at 9754 m (32,000 ft) and closing on the fighter-bombers he was protecting. He turned to meet the threat, joined only by his wingman, intending to break up the attacking force before it could get among the bombers. Davis destroyed two of the MiGs, to take his score to 12 victories, and was closing on a third when his Sabre was hit and crashed into a mountainside near Tong Dang-dong. Davis was awarded a posthumous Medal of Honor for his heroic action.

On 23 February, Maj. William T. Whisner, of the 51st FIW, destroyed his fifth MiG to become the Wing's first ace and the seventh of the Korean War. Whisner, who commanded the 25th

FIS 'Assam Dragons', had scored 15.5 confirmed victories in World War Two.

Another reason for the dearth of 4th FIW victories during January–February was the lack of sufficient numbers of trained combat pilots to replace the experienced 4th FIW career pilots, many of whom had by now been rotated home after 100 missions. The situation only improved in March when increased serviceability permitted more combat operations, and the Sabre wings began to receive young fighter pilots fresh from training in the US. By now FEAF had 165 Sabres in the Far East and 127 of these were available for combat in Korea, but a lack of spares and poor maintenance grounded many Sabres. On average 45 per cent of the Sabres were unserviceable, 16.6 per cent because of lack of spares and 25.9 per cent because of maintenance problems. External fuel tanks too were in very short supply and Sabre pilots were forced to fly combat patrols with only one tank. Further supplies of fuel tanks were flown to the combat area direct from the contractors in the USA, but in February the 4th and 51st FIWs had to reduce their combat sorties to a minimum. Air Materiel Command launched a crash programme called Peter Rabbit to raise stocks of spares to an acceptable level and by April 1952 the unserviceability rate for lack of spares was down to 2.4 per cent.

New Sabre tactics

In March 1952, the Sabre formations began entering the combat area stacked down from 12192 m (40,000 ft) to obtain a better chance of engaging the MiG-15s, and pilots claimed 39 MiGs for the loss of only three Sabres and two F-84s. April was even better, with 44 MiGs claimed destroyed for the loss of four F-86s and one F-80, although the total would have been higher if the American pilots had been allowed to cross the Yalu into Manchuria. Instead, they were restricted to making strafing and bombing attacks on Communist airfields south of the

Lt-Col George I. 'Shakey' Ruddell (right), CO, 39th FIS, was photographed at K-13 on 18 May 1953 after becoming the 30th ace of the Korean War. Ruddell ended the war with eight victories. *(via Larry Davis)*

Yalu. On 22 April Capt. Iven C. 'Kinch' Kincheloe, 25th FIS, 51st FIW, who had become an ace on 6 April, and Maj. Elmer W. Harris, destroyed two Yak-9s at Sinuiju airfield in strafing runs. They returned on 4 May to strafe 24 Yak-9s and five were destroyed in the attack, Kincheloe accounting for three of them (in September 1956 Kincheloe took the experimental Bell X-2 to Mach 2.93; he was killed flying a Lockheed F-104A Starfighter at Edwards AFB on 26 July 1958).

Others who became aces in April were Capts Robert H. Moore and Robert J. Love and Maj. Bill Wescott. In May Capt. Robert T. 'Cowboy' Latshaw, Maj. Donald E. 'Bunny' Adams, Lt James H. Kasler and Col Harrison R. Thyng also became aces and James Kasler later increased his score to six. On 13 May 4th FIW Sabres, each carrying two 1,000-lb (454-kg) bombs below their wings, made their first dive-bombing attack on Sinuiju. In another attack, on the railway yards at Kumu-ri, Col Walker H. 'Bud' Mahurin, 4th FIW CO, was shot down by flak and taken prisoner. (This was the second time in his career that Mahurin had been shot down, the first being on 27 March 1944, when he managed to evade capture. Flying Republic P-47 Thunderbolts in the 63rd FS, 56th FG, he was credited with 19.75 German aircraft destroyed and he also shot down a Japanese bomber in the Pacific). Mahurin was credited with 3.5 MiGs, one probable and one damaged, at the time of his capture in Korea. He was not released until September 1953.

On 14 May on his first encounter with an enemy aircraft, Lt Edwin 'Buzz' Aldrin of the 51st FIW shot down a MiG. (He ended his tour with 66 missions after shooting down another aircraft and damaging a third. In the late 1950s Aldrin became an astronaut and on 20 July 1969 he became the second man, after Neil Armstrong, to walk on the surface of the moon in the climax to the *Apollo 11* lunar landing). In May 1952 the enemy lost 27 MiGs while the Americans lost five Sabres, three F-84s and one North American F-51 Mustang in combat during

a record 5,190 sorties. This total remained unsurpassed when hostilities ended.

Though an expansion programme had been under way to increase Sabre strength from the five wings on hand in June 1951, due to the time it took to increase production, just seven Sabre wings were active in June 1952. Only two of these were allotted to Korea because of USAF commitments but early in June, the 51st FIW was strengthened with the arrival of the 39th FIS, 18th FIW, equipped with the latest F-86F model.

Improved F-86F arrives

The new fighter showed considerable improvement over the E model. Sabre pilots had reported that intermittent opening of wing slats on the F-86E caused them gun-sighting problems during combat. The wing slats were omitted on the F-86F version and a new wing leading edge, extended by 22.86 cm (9 in), was added to improve manoeuvrability at high altitudes. In June, Sabre pilots claimed 20 MiGs destroyed for the loss of three F-86s. Lt James F. Low of the 4th FIW became an ace during the month. On 4 July 50 MiGs crossed the Yalu and UN pilots claimed 13 destroyed for the loss of two Sabres. That month 19 MiGs were shot down while four Sabres failed to return. Capt. Clifford D. Jolley, 335th FIS, 4th FIW shot down a MiG-15 on the 8th to take his score to five. In August, 35 MiGs were shot down, including six which were destroyed in a battle between 35 Sabres and 52 MiGs on the 6th.

In September the 335th FIS, 4th FIW received F-86Fs to replace its F-86Es. One of the heaviest battles of the year took place on 4 September, when 13 MiGs were shot down in 17 separate air battles for the loss of four Sabres. Capt. Frederick C. 'Boots' Blesse of the 334th FIS, 4th FIW destroyed one of the MiGs to become an ace. He notched up a further four victories before the end of the month and one more in October to take his score to ten confirmed victories. On 9 September, F-84s attacked the North Korean Military Academy at Sakchu and some of a force of 175 MiGs broke through the Sabre screen and

shot down three fighter-bombers. Two more Thunderjets and six Sabres were destroyed in air combat during the rest of the month but the UN pilots claimed a record total of 63 MiGs. On 21 September Capt. Robinson Risner destroyed his fifth MiG, near Sinuiju, to become the Korean War's 20th jet ace. The heavy losses made the Communist pilots more cautious during October,

F-86F-5 51-2940 *MIG MAD MAVIS* of the 39th FIS, 51st FIW, has seven stars denoting MiG kills in Korea. It was flown by Lt-Col George I. 'Shakey' Ruddell, CO, 39th FIS. *(via Robert Jackson)*

Captain James J. Jabara of the 334th FIS, 4th FIW, was at the controls of F-86F-1 51-2894, for this summer 1953 photograph. *(via Larry Davis)*

when the Sabres destroyed 27 MiGs for the loss of four F-86s and one Thunderjet. In November only the more experienced MiG pilots dived down from the safety of numbers at 12192 m (40,000 ft) to take on the American formations. On occasions, smaller formations of up to 24

Capt. Ralph S. Parr, of the 335th FIS, 4th FIW, was photographed in F-86F-30 52-4778 *Vent de la Mort/Barb* over North Korea in the late summer of 1953. On 27 July 1953, the day that an Armistice was signed ending the war in Korea, Parr destroyed an Il-12 transport. It was the last aircraft to fall in the Korean War and Parr's tenth victory. *(via Larry Davis)*

MiGs took on flights of four Sabres and tried to box them in. The Americans responded by increasing their flights to six or eight aircraft, with the higher-performance F-86Fs operating at 12192 m (40,000 ft) and covering the more vulnerable F-86Es at lower altitudes. The new tactics worked, for only four Sabres were lost in combat while 28 MiGs were destroyed in the air.

In November three more Sabre pilots – Col Royal N. 'The King' Baker, CO, 4th FIW, Capt. Leonard W. Lilley of the 334th FIS, 4th FIW, and Capt. Cecil G. Foster of the 16th FIS, 51st FIW, became aces during the month. In December 1952 the MiG pilots ignored the Sabre screen and headed south to the Chongchon River where they ambushed Sabres returning home short of fuel. At least four Sabre pilots had to bale out when their fuel was exhausted, but only two F-86s were lost in air combat that month, while UN pilots claimed 28 MiGs. In January 1953, 37 MiGs were destroyed. Capts Dolphin D. Overton, III of the 16th FIS and Harold E. Fischer of the 39th FIS, 51st FIW became aces during

January. Overton destroyed five MiGs and damaged another in just four missions. On 18 February four Sabres attacked 48 MiGs, shooting down two and forcing two more to spin and crash. In all, 25 MiGs and four Sabres were lost in combat during February. On 3 February 1953 at Osan-ni airfield the 18th FBW's three squadrons, the 12th and the 67th, and No. 2 Squadron, South African Air Force (SAAF), finally began conversion from Mustangs to F-86Fs. The Wing flew its first combat mission with F-86Fs on 25 February on a fighter sweep to the Yalu. Col Frank S. Perego, 18th FIW commander, was not satisfied with many of the ex-Mustang pilots and he reassigned them to other 5th AF units, replacing them with pilots from the US, but the unit was fully operational by early April. On 22 February the 35th and 36th Fighter-Bomber Squadrons (FBSs), 8th Fighter-Bomber Wing (FBW) at Suwon also began conversion, from F-80Cs to F-86F Sabres. The 8th FBW flew its first Sabre mission on 7 April, when four Sabres joined a fighter sweep to MiG Alley.

The 80th FIS began conversion to the F-86F on 30 April. On 13 April the F-86Fs were fitted with bomb shackles, special 200-US gal (757-litre) drop tanks and a gun/bomb/rocket sight and the 8th FIW flew the first F-86F fighter-bomber mission. It was followed by the 18th FIW on 14 April.

Mounting MiG losses

In March 1953, 34 MiGs and three Sabres were lost and in April the Sabres claimed 27 MiGs destroyed for the loss of just four F-86s. A fifth Sabre was shot down by flak. On 12 April Capt. Joe McConnell of the 16th FIS, 51st FIW was forced to eject to safety after his Sabre was badly hit, but a helicopter of the 3rd Air Rescue Squadron picked him up. He was back in action within 24 hours to shoot down his ninth MiG. His tenth victory came on 24 April, putting him level with Capt. Manuel J. 'Pete' Fernandez of the 334th FIS, 4th FIW. On 27 April, Fernandez shot down his 11th enemy fighter to lead the table of Korean War aces. May 1953 was a highly successful month for marauding UN fighter pilots who no longer faced the Soviet pilots who had been flying the MiGs in combat. From 8–31 May, some 1,507 MiGs were sighted and in engagements with 537 of them, 56 were shot down. On 18 May, Capt. Joseph McConnell scored his 16th and final victory when he destroyed three MiGs. Both he and Fernandez, who finished the war with 14.5 kills, were pulled out of the war on 19 May and sent Stateside. McConnell's score remained unbeaten, making him the leading ace of the Korean War.

In June, 77 enemy fighters were shot down, 11 probably destroyed and 41 damaged, without loss to UN forces. Sixteen of the victories were claimed on one day, 31 June – a new record. The June victories saw five new aces: Lt-Col Vermont Garrison, Capt. Lonnie R. Moore and Capt. Ralph S. Parr of the 335th FIS, 4th FIW and Col Robert P. Baldwin and Lt Henry 'Hank' Buttelmann of the 51st FIW. In July, the Sabres alone shot down 32 enemy fighters. On 11 July Maj. John F. Bolt, USMC, flying in the 25th FIS,

51st FIW, shot down his fifth and sixth MiGs to become the only USMC ace of the Korean War. Maj. John H. Glenn, Jr, USMC, another exchange pilot, scored three victories while with the squadron. The 'MiG Mad Marine', as he was known, later became an astronaut and was the first American in space.

Maj. James Jabara had returned to the USA with six kills, before returning to combat in January 1953. By 26 May, he had downed three MiGs and on 10 June the 4th FIW ace shot down his tenth and 11th MiGs. He added three more that month before claiming his 15th and final MiG victim on 15 July. Jabara's 15 victories put him into second place behind McConnell. On 19th and 20th two more 4th FIW pilots, Capt. Clyde A. Curtin and Maj. Stephen L. Bettinger, also became jet aces. Bettinger was the 39th and last jet ace of the war, but it was several months before his status could be confirmed because he was shot down and captured, and the UN kept his kills secret until his safe repatriation. At 17:00 hours on 22 July, three Sabres of the 51st FIW, led by Lt Sam P. Young, entered MiG Alley at 10668 m (35,000 ft) on an offensive patrol. It was Young's 35th mission and he had yet to fire his guns in anger, when ahead and below he saw four MiG-15s sweep across his path at right angles. Young dived down and destroyed the No. 4 with a long burst of fire. It was the last time that the Sabre and MiG met in combat. On 27 July an Armistice was signed. That same day Capt. Ralph S. Parr, of the 335th FIS, 4th FIW destroyed an Ilyushin Il-12 'Coach' transport. It was the last aircraft to fall in the Korean War and Parr's tenth victory.

Final analysis

On 29 July 1953, a 5th AF communiqué stated that the war had cost 58 F-86s shot down and that 808 MiGs had been destroyed by Sabres; a ratio of 13.79:1. These totals had been achieved despite the Sabre pilots having to operate over enemy territory the whole time and at the extreme limit of their aircraft's range, which restricted patrol time along the Yalu to under

15 minutes. The MiG pilots chose the time and place for action and broke off combat when it suited them. So how did the Sabre pilots (who were usually vastly outnumbered in combat) manage to knock down so many MiGs for a relatively low combat loss rate of F-86s? The UN airmen were experienced pilots while the majority of Chinese flyers were greenhorns by comparison. Often they would panic and fire wildly. Many put their fighters into unnecessary spins and in the last resort some chose to eject rather than stay and fight. (In the last four months of 1952 a fifth of the Sabre victories were achieved without the F-86 pilots firing their guns! Some 32 MiGs were seen to snap suddenly into spins while manoeuvring to escape and eight pilots ejected, 22 crashing with their aircraft). A FEAF study in March 1954 declared that: '...68 per cent of pilots who had destroyed MiGs were over 28 years old, while 67 per cent of the pilots who had scored no kills were less than 25 years old. Pilots with MiG kills had flown an average of 18 missions in World War Two, while pilots with no kills had flown an

average of four missions in World War Two. Some 810 enemy aircraft were claimed destroyed by Sabres and the 39 Sabre pilots who became jet air aces destroyed 305.5 aircraft. Whether or not a pilot was flying as an element leader and the conditions under which he sighted MiGs affected his chances for scoring victories, but the more experienced pilots apparently had the best chance for shooting down the enemy...'.

Sources since have lowered the number of MiGs shot down by Sabres to as low as 379 and Russian archives admit the loss of 345 Soviet-piloted MiG-15s, while other sources reveal that the Sabre:MiG kill ratio was between 10.32:1 and 14:1, with the higher score the more likely. Whatever the scores, unquestionably, without the Sabre, the Communists would have gained air superiority in Korea and the war would have been lost.

Harmonising the Sabre's guns at K-55 Osan in 1953. Normally, the Brownings were harmonised so that the six trajectories converged 1,000 ft ahead of the F-86. *(Bill Grover via Larry Davis)*

3. Flying the F-86 Sabre

The Engagement
1/Lt (later Lt Col) Reg Adams

'It was a beautiful flying day in Korea with unlimited visibility as the 39th Squadron launched a full-blown Yalu Sweep on 19 June 1953. No less than 48 Sabres from the 16th, 25th and 39th Squadrons were lined up on the runway at K-13 [Suwon] in central South Korea. The air-to-ground boys across the field [the 8th FBW] were scheduled to launch soon afterward. Leading my flight was Colonel (later General) George "Shakey" Ruddell, 39th FIS commander. I was flying No. 4 as wingman – CHAPTER THREE – to Lt. Wade "Killer" Kilbride. We were COBRA FLIGHT, which coincidentally was also the emblem of the 39th Squadron. Flying with the squadron commander was not exactly every pilot's dream, because he was always the most demanding. We also suspected that the engine in his F-86 was a little 'souped up', so to speak. The only setting that Col Ruddell knew on the throttle was full forward from take-off to landing.

'Our mission was to intercept any MiGs attempting to cross the Yalu River and attack the F-84 and F-86F fighter-bombers that were working targets in North Korea. Soon after arriving at our patrol station on the Yalu River, we spotted six MiGs in formation attempting to slip into North Korea at low altitude. Col Ruddell immediately began a dive, which put us right on top of and directly behind the MiG formation, i.e. the perfect

'bounce' from 6 o'clock high. The Colonel and his wingman took on the MiG leader. Kilbride set his sights on the leader of the second element. The other two MiGs broke their formation and disappeared for the moment. Though we lost sight of Ruddell, he eventually shot down the MiG that he had engaged. He was already an ace and this was his 8th victory of the war.

'Kilbride, my leader, engaged his MiG in a tight turn, firing continuously and scoring numerous hits on the Russian fighter. I attempted to stay on his wing, protecting his tail and watching the MiG Wade had staked out. Thank God for the g-suit, because I was holding a constant 4 g trying to stay with Wade and the

1/Lt Reg W. Adams of the 39th FIS, 51st FIW, poses on the wing of his Sabre *Dorothy Ann*, which he named in honour of his wife. (*Dorothy Adams*)

1/Lt Reg W. Adams of the 39th FIS, 51st FIW, dressed for combat. *(Dorothy Adams)*

MiG in the turns. In the course of all this action, the enemy wingman appeared on my left side attempting to get into a firing position on Kilbride. As the MiG pulled up on my left, I held my *g* forces until I felt that it was time for me to do something to prevent his firing on Wade.

'I relaxed just enough stick pressure to put me in position to fire. My .50 calibre tracers laced through the canopy of the MiG, which immediately did a lazy roll and headed for the ground. In spite of my gun camera film confirming this part of the action, I didn't see any type of explosion. I suspect that my bullets may have killed the MiG pilot, as my tracers penetrated the fuselage where the MiG had very little armour protection.

'However, the intelligence people would not confirm the victory. Many times I have wondered if I should have broken off and followed that MiG down to get the confirmation. But, needless to say, as a wingman I was committed to staying with my leader and protecting his tail. Shortly thereafter, Kilbride 'fired out' [expended all his ammunition] and called on me to continue the engagement with 'his' MiG. I pulled in behind the MiG Wade had been firing on. The MiG pilot, thinking the engagement was over, rolled out straight and level, turned north and headed for the Yalu and safety. I very deliberately pulled in right behind the MiG, put my pipper on his tailpipe, and almost counted a kill. Suddenly I noticed what appeared to be flaming ping-pong balls floating past my Sabre. Cannon shells! Really big 37-mm cannon shells! I heard a frantic call from Wade, "COBRA 4, break right now!" I had no choice but to break off from a certain victory and head for home.

'Later, Wade and I determined that the two MiGs we thought had abandoned the fray after our initial bounce had decided to come back and help their comrades. We also figured they had received a bit of "encouragement" from the MiG that Wade and I were firing on, i.e. Chinese for "Get these guys off my tail!" My hard right break saved my life, as the MiGs didn't give chase, which allowed us to return to Suwon safely. There were a lot of hairy stories floating around the bar that night because we, the 51st Group, had several confirmed kills that day. Kilbride bought me a drink!'

USAF Fighter Operations
Group Captain R. J. F. 'Dickie' Dickinson, AFC

'Volunteers were called from experienced day-fighter pilots in RAF Fighter Command then filtered down to a number agreed with the USAF. Air Marshal Sir Basil Embry at Headquarters Fighter Command then interviewed the selected pilots in November 1952, before flying out to Nellis AFB in January 1953. The RAF did not have a suitable fighter at that time to match the MiG-15. Magnificent though the Meteor F.Mk 8 was, it could not cope with high Mach numbers and the very high ceiling of the MiG. In February 1953, I, and a small group of pilots, commenced a very intensive six weeks' conversion to the F-86E: dog-fighting, tail chasing, formation flying, dive bombing, rocketing, air-to-air flag firing, night flying and simulated sweeps of four aircraft, the latter with experienced bouncers. The staff pilots on all these training sorties were all highly experienced and most had completed 100

Left: Flt Lt R. J. F. Dickinson, RAF, of the 25th FIS, 51st FIW, successfully landed F-86E-10 51-2825 *Bella* on the beach at Pyong-Yang-Do on 15 July 1953, after a mid-air collision with Lt Aaron. *(Grp Capt. Dickinson)*

Flt Lt (later Grp Capt. R. J. F. 'Dickie' Dickinson, AFC) was an RAF exchange pilot flying F-86Es with the 25th FIS, 51st FIW in Korea. He was photographed on 2 May 1953. He was one of 21 RAF pilots who operated with USAF fighter squadrons in Korea. Five MiG-15s were confirmed shot down by four RAF pilots, including Dickinson, who got his victory on 18 June 1953, flying F-86E 52-2882. He was awarded an American DFC. Ten RAF pilots were lost while attached to the USAF and RAAF. *(Grp Capt. Dickinson)*

Flt Lt John H. J. Lovell, RAF, of the 25th FIS, 51st FIW is shown in the cockpit of shark-toothed F-86E-10 51-2825 *Bella*, which was assigned to Capt. George Howell. Lovell was credited with one MiG kill in Korea. *(George Howell via Larry Davis)*

missions on F-86s in Korea. They were all very good and put us though our paces. After completion of the course it was off to Japan, where we were allocated to our future squadrons in Korea. I was posted to the 25th FIS at Suwon, where there were already two RAF pilots – Flt Lt Jock Maitland and Flt Lt John Lovell. On arrival, we were made very welcome and commenced our introductory flying programme before being let loose on Yalu sweeps. Amongst my mentors for my first few sorties were Lovell and Maitland who gave me an excellent initiation into the local inhospitable terrain as well as more dog-fighting, tail chasing and sticking in as a wingman through every type of manoeuvre.

F-86 Missions

'The prime purposes of the F-86 were to maintain air superiority over the Korean peninsula, and to destroy as many MiG-15s as possible in aerial combat. Air superiority involved escorting ground-attack and recce aircraft. My first Yalu sweep was on 13 May, involving 48 Sabres, but no MiGs were sighted in our sector. This was repeated on 15 May, but although many MiGs were sighted they were all well to the east of our area and flying above 50,000 ft [15240 m] so no contact was made. Things were to change on 16 May when another Yalu sweep was planned. I was to fly No. 2 to Jock Maitland; we were part of 24 aircraft from our Squadron. We climbed up to 35,000 ft [10668 m] and toggled off our drop tanks – which we did on every mission. As we got to about 50 miles [80 km] from the Yalu we checked our guns. My guns worked but my gunsight was u/s [unserviceable]. I had a quiet word with Jock who said he had fuel feed problems and in view of this we were going to abort and return to base. He then called a "turnabout left". When we had passed through 90° I saw four MiGs barrelling down on us from 8-o'clock high. I immediately called "break left". As I did this I saw three of the

Above: USAF and RAF exchange pilots of the 25th FIS 'Assam Dragons', 51st FIW, pose on the wing of an F-86E at a forward air base in Korea in 1953.

Right: Flt Lt R. J. F. Dickinson, RAF, of the 25th FIS, 51st FIW, holds a parasol in the cockpit of an F-86E in Korea in 1953. *(both Grp Capt. Dickinson)*

MiGs shoot past my tail unable to hold the turn, but the fourth MiG, pulling like hell, managed to get almost between Jock and myself. I thought he was going to collide with me so I slapped out my airbrakes and throttled back. By this time the MiG was filling my windscreen at about 50 yards [46 m]. I pulled my trigger – still no gunsight but fired up his tailpipe aiming on my tracer. I got a number of hits, upon which he broke hard left and dived steeply in a left spiral. I followed him down, still firing. I was so stunned by everything

that had happened in seconds, that I failed to tell my leader what was happening or to advise him to come out of his turn and cover me.

'Whilst still firing at the MiG who by now was really smoking, I heard Jock call me to break left since there was a MiG firing at me. I then broke off my attack and turned hard left. A quick look all around confirmed there was no MiG behind me and that I was alone. My MiG was still spiralling down vertically, still smoking heavily. By chance I saw Jock orbiting to the south of me

Col Harrison R. Thyng, a World War Two ace with 11 kills, as CO, 4th FIW, attained Korean War ace status flying F-86E-1 50-623 on 20 May 1952. (via Larry Davis)

10,000 ft [3048 m] above. I joined up with him and we returned to Suwon. On our debrief Jock quite rightly chewed me up for not keeping him informed and at the same time told me that after he had seen me disappear down, he thought it was a MiG on my tail, whilst in fact it was me on the MiG's tail! I learnt my lesson from that engagement. In a lame excuse it was the first MiG I had ever seen close up – so nearly a disastrous mid-air collision and a lost opportunity to get a positive kill. I was credited with a probable.

'I only fired my guns twice more in anger; once when flying wingman to my Squadron Commander, Major Giraudo. He was chasing a MiG about 1,500 yards [1372 m] ahead when two MiGs at 7-o'clock high bounced us. I called a break left, but there was no reaction from my boss. I could see the lead MiG firing and his 37-mm cannon shells streaking towards my

leader like flaming red tennis balls. He then switched over to me. I broke hard left then reversed on the MiG as he dived down and back to the Yalu River. I opened fire and got in a long burst, but he was, by now, out of range.

'On 18 June 1952 I was briefed to fly No. 3 in a formation of four aircraft led by Col Robert P. Baldwin, our Group Commander. My No. 2 had to abort on the R/W so the boss said, "Okay, let's go as a three". We climbed up through cloud and then descended back though cloud to 20,000 ft [6096 m] over the Yalu. As we came out into the clear, I saw two MiGs right behind us at about 1,000 yards [914 m]. I called break left and one MiG flew past my tail and another followed my leader in a very hard turn. I managed to get on his tail and started firing. He turned on to a north heading, before straightening out and starting to burn. My leader was then covering me. I fired every round in my aircraft; the MiG started to spiral down and crashed near Oick-Tong. This was the last time I fired my guns in anger. I was credited with a kill.

'For the next three weeks I was involved in escorting F-84s on interdiction missions and US Navy Banshees on photo sorties. On 15 July, I was detailed to lead 4 aircraft on a low-level recce mission of the Antung MiG airfield complex at 1,500 ft [457 m]. We flew up the west coast at 40,000 ft [12192 m] until over the mouth of the Yalu, then put our noses down in "Finger Four" formation at about 0.95 Mach, arriving near Antung doing over 600 kt [1112 km/h; 691 mph]. I had staggered my four at varying heights to avoid the flak, which was intense. I had binoculars round my neck to look out for new revetments on some of the airfields planned to receive [Ilyushin] Il-28 ['Beagle'] light bombers. Lt Aaron, my wingman, was getting rather excited, calling out very accurate flak near Taku-Shan. I then turned slightly right. As I did this everything turned black, followed by a very loud bang, after which my aircraft turned on its back. I then realised my wingman had hit me as he was trying to cross under me. My aircraft [51-2825 *Bella*] decelerated from 610 kt [1130 km/h; 702 mph] to 180 kt [334 km/h; 208 mph]; but still continued to fly. I saw my wingman climbing away and disappear, heading south. (Aaron got back home with a missing tailplane). My Numbers 3 and 4 had to return to base because of fuel shortage. My mayday had brought a few sympathetic calls of good luck, etc. To my surprise, my aircraft continued to fly, although it would go no faster than 175 kt [324 mph; 201 mph] at full throttle. As I crossed the coast, still with no MiGs around and no flak, I made preparations to eject. I knew there was a small island (Pyong-Yang-Do) about 50 miles [31 km] to the south west, so I plodded down the coast well out to sea. After a while I saw the Island with its long beach, which I knew was manned by US Marines.

'I once again prepared to eject, but my aircraft still handled okay and the engine kept going. To cut a long story short, I managed a successful wheels-down [landing] on the hard sand. As I came to a stop, a few Marines emerged from dugouts near the beach and greeted me. I was returned to Suwon by light aircraft the next day, and back on ops the day after. My wingman had returned safely without knowing he had lost half a tailplane. The war ended on 27 July, 12 days after this incident, so I suppose I was very fortunate in not ending up in a North Korean PoW [prisoner-of-war] camp.

'I believe it is worth recalling that I was leading eight Sabres on 12 August (after the war) up the west coast of Korea on an early morning mission, well out to sea, when we were recalled because of bad weather approaching our base. On arrival overhead it was obvious that Suwon

Captain James J. Jabara of the 334th FIS, 4th FIW is shown at K-14 Kimpo in 1953. Jabara returned to the USA with six kills, before returning to combat in January 1953. He claimed his 15th and final MiG on 15 July, which put him into second place behind Joseph McConnell at the war's end. *(via Larry Davis)*

was weathered out and that we had no diversions. I had advised the operations staff that this might happen, but was firmly told to get airborne. After an abortive approach to Suwon, I climbed back through terrible weather and rejoined the rest of my formation, which I had left at altitude, and carted them off to Pyong-Yang-Do Island, where all except my wingman, who had damaged his undercarriage, completed successful wheels-down landings on the beach. A Dakota flew up that afternoon with fuel and groundcrew and we all flew back to Suwon the next day, with the exception of my No. 2, who returned in the [Douglas] Dakota.

'At the same time as this drama was unfolding, Jock Maitland was leading eight Sabres on a similar mission up the east coast of Korea and was faced with an identical problem. After the weather recall he saw a small gap in the clouds with an airstrip visible. He dived his formation, and under very heavy rain, landed them all safely on a South Korean airfield with no damage other than some aircraft sliding off the end of the runway.

'Another interesting mission took place after the war on 28 August, when I was included in a 12 aircraft escort for two Banshee recce aircraft to photograph some Soviet airfields close to Vladivostock. We put on 200-US gal [757-litre] drop tanks, joined up with the two recce aircraft and headed across Korea to the east coast, then up to Vladivostock at altitude. The Banshees dropped into 1,000-yard [914-m] lines astern at about 25,000 ft [7620 m] for the photo run and we flew top cover at 35,000 ft [10668 m]. We expected swarms of MiGs and possibly flak, but surprisingly saw neither. The mission was, of course, completed out to sea and took 2 hours.

The F-86

'The Sabre was a truly magnificent aircraft and a delight to fly. Sitting up in that excellently positioned cockpit I felt I was king of the castle, and would survive in any situation and perhaps immodestly get a few more MiGs if the war had continued a little longer. One of its greatest assets was the all-flying tail. One could take it up to 40,000 ft [12192 m], turn it on its back and pull through, still maintaining full elevator control with the Mach indicator hovering just over Mach 1. The bubble canopy, with its superb rear view, was an invaluable asset. The ailerons were crisp at all altitudes and speeds. It was very docile in the circuit and landing. Its constraints lay in its slight lack of engine power. As far as armament was concerned, the 0.5-in [12.7-mm] machine guns did very well, but two or four 20-mm cannons would have been much more lethal – just recalling the later [Supermarine] Spitfires and [Messerschmitt] Bf 109s.

Tactics

'We would often start off a sweep with 48 aircraft sub-divided into fours. Once MiGs were sighted, the fours would spread out until close contact was made, then split into pairs during combat, covering each other if possible. It was the wingmen's task to cover their leaders and only shoot in extreme circumstances. On a number of occasions, we sighted MiGs flying in trains of about 16 aircraft at heights of above 50,000 ft [15240 m]. They would sometimes detach four or more aircraft and dive down on us, have a quick burst – often out of range – and then 'high tail' it off to the north for the sanctuary of China. On these occasions we had few alternatives other than wait until we were bounced, then break at the appropriate time, reverse quickly as they shot past our tails and have a going-away shot. Once they turned and tried to tangle with us, they usually lost out and were shot down.

'One of our biggest problems was fuel. We had to cover 200 miles [322 km] of hostile territory to get to the combat area of the Yalu; this left us with only 15 minutes for combat before "bingo" time and the return 200 miles to Suwon.

'Our kill ratio was about 12 to 1; the total number of MiGs destroyed in combat was about 790. I finally completed 42 missions before the war ended.

'The quality of pilots was excellent and similar to those I had known on RAF squadrons but this, of course, was the testing of USAF pilots in a real hot war. I would like to pay tribute to those Royal Navy, Army Air Corps, and RAF pilots who lost their lives in Korea and those who suffered as PoWs of a voracious and cruel enemy. With many thanks to the USA and the USAF, who made us so welcome as "Brothers in Arms".'

Indo-Pakistani War, 1965
Group Captain Mohammed Shaukat-ul-Islam, Pakistan Air Force

'In November 1964 I was posted to 11(F) Squadron, commanded by Sqn Ldr M. M. Alam at Sargodha. I became operational in August 1965 and was allowed to take part in the 6–23 September 1965 war with India. I considered myself very lucky to have taken part in the war as a Flying Officer (Lt), with only about 80 hours on the F-86F and with a grand total of about 400 hours. At the outbreak of the war, No. 11 Sqn was tasked to carry out a dawn strike against the Indian army in the Chamb-

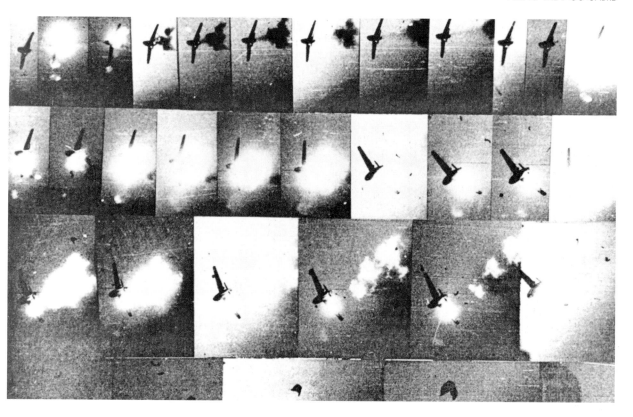

Jurian sector with two formations of eight F-86 aircraft. Each aircraft carried 32 5.75-in [14.6-cm] rockets and 1,800 rounds of 0.5-in [12.7-mm] ammunition. We exhausted all the weapons on the Indian army convoy and returned to Sargodha safely. As it was a surprise dawn strike, we faced only small arms fire from the enemy. By the time I landed and cleared the runway, my aircraft flamed out because of shortage of fuel.

'On 9 September, four F-86Fs were tasked to provide a low-level escort mission for three [Martin] B-57 [Canberra] bombers attacking a train carrying ammunition at Gadro. The bombers carried out four attacks each and all seven aircraft remained within heavy 'ack-ack' fire [flak] for about 15 minutes. All aircraft exited low level after the successful delivery of their weapons. The three bombers recovered at Peshawar and we four fighters came back to Sargodha safe and sound. It was my first

A Pakistan Air Force (PAF) F-86F explodes under the gunfire of an Indian Air Force (IAF) Hawker Hunter during the Indo-Pakistan War of 1965. The PAF lost 13 F-86Fs, seven of them in air combat. *(via Robert Jackson)*

experience of remaining within such heavy anti-aircraft fire for such a long time.

'On 11 September, in a formation of four F-86Fs, I took part in a daytime escort mission to give air protection to a train carrying ammunition from Lahore to the Sialkot sector. It might sound very easy, but to give protection to such a slow moving train from so fast moving an aircraft, at low level, by four Sabres for such a long time, was very demanding. However, the train reached its destination and off loaded its cargo.

Air combat with the Hunter
'On 16 September I took off from Sargodha as Sqn Ldr M. M. Alam's wingman, to carry out a high-level offensive patrol mission deep inside

Indian territory. We were flying in battle formation at 23,000 ft [7010 m] between two Indian air bases, Halwara and Adampur. The aim was to invite the Indian fighters to come and fight with us. We could undertake such a venture because by then the Pakistan Air Force [PAF] had established air superiority over the Indian Air Force [IAF]. It was at about 2 pm, with clear blue skies, when our ground controller at a radar station transmitted that two IAF [Hawker] Hunters had taken off from Halwara and were approaching to intercept us. When they came in sight we jettisoned our drop tanks and entered into close air combat. The air battle became intense and under such high-*g* manoeuvres I could not stay on the tail of my leader. As it turned out, my leader shot the No. 2 of the other formation and their leader shot me. My aircraft caught fire and I ejected through the shattered canopy at about 12,000 ft [3658 m]. I lost consciousness for a couple of seconds, and by the time I got my senses back I was floating in the air and the small parachute was pulling out the bigger one. As I settled down in my parachute I saw a Hunter streaming fuel and crash with a big explosion. The Hunter pilot was shot in the cockpit. When I looked down to locate my probable landing spot, I noticed with horror that a man in uniform was pointing a 0.303 [7.7-mm] rifle at me and a civilian was aiming a double-barrelled shotgun. I heard three

shots and within seconds my feet touched the ground. I got up, released the parachute and was surrounded by a crowd of people. The name of the place was Taran Taran. The local police rescued me from the crowd and took me quickly to a nearby police station and then to a hospital. I was bleeding profusely from my back. A doctor operated on me and showed me the 0.303 bullet he had taken out of my back. Next day I was taken to IAF base, Adampur and flown by An-32 to Delhi and admitted in the CMH. The cease-fire was declared on 23 September when I was still in the CMH. Later, I joined a pilot and navigator of a B-57 which was shot down by AA fire on 15 September in a night raid over the Adampur base. We three returned to Pakistan after being released in a prisoner exchange in February 1966.

'In the war of 1965 I flew a total of 19 missions, including day and night air defence missions up to 16 September. The story of my time as a PoW is a different chapter of my life. However, I can say that the IAF treated me very well. In later days, when I joined the Bangladesh air force in 1972, I had the opportunity to visit the IAF as an official guest and met many friends with whom I had come into contact as a PoW. After returning from India I was posted back to No. 11 Sqn. From then on it became my passion to be a master in air combat. In my later days I could fly the F-86 like a toy and used to manoeuvre it to its design limits. In the early 1960s we used to comment that a pilot who had not flown the F-86 had not enjoyed the charm of fighter flying. I was later posted to No. 14 Sqn, at Dhaka, and No. 26 Sqn, at Peshawar, where I continued flying the F-86F. In 1968, the PAF introduced the F-86E and soon it became a very popular fighter aircraft. I continued flying both models until 1970, logging about 1,200 hours on the F-86F and E combined. In total, I flew 13 types of aircraft in my career, including the [Mikoyan-Gurevich] MiG-21MF ['Fishbed-J'] and the [Northrop] F-5.'

Mohammed Shaukat-ul-Islam of the Pakistan Air Force.
(Grp Capt. Mohammed Shaukat-ul-Islam)

4. Variants

RF-86A/F

In Korea the RF-80A and the North American RB-45C Tornado reconnaissance aircraft were found to be too slow to operate in MiG Alley without escort and since the RF-84F Thunderstreak was not yet available, in October 1951 six F-86A-5 Sabre aircraft were set aside for conversion to RF-86A configuration. A compartment below the cockpit was enlarged and fitted with constant-temperature conditioning for a forward oblique 24-in (610-mm) K-22 camera and a pair of 20-in (508-cm) K-24 cameras mounted lengthwise with a mirror arrangement to provide vertical

F-86A-5s 48-201, 48-244 and 48-240, of the 3596th CCTS 'Cadillac Flight'. *(via Larry Davis)*

F-86E-1 50-599 was one of 60 (50-579/638) from the first production order for 111 NA-170 aircraft, delivered to the USAF from 9 February 1951. *(via Robert Jackson)*

coverage. Five RF-86A Sabres, each with their distinctive cheek fairings, were issued to the 15th TRS, 67th Wing in Korea. The first time one was attacked, in February 1952, it turned towards the MiG-15, which was unable to turn in time to make another attack. However, the vertical camera was too slow in its intervalometer operation to be suitable. One RF-86A was lost in June 1952. At least five more F-86A 'Ashtray' conversions were carried out and in 1953, in 'Modification Haymaker', several F-86F-30 Sabres were equipped with two K-22 and one K-17 camera. These Sabres did not need cheek fairings because the cameras were mounted horizontally. After Korea the California Air

National Guard (ANG) operated one RF-86A, ten went to the Republic of Korea Air Force (RoKAF) and seven RF-86Fs were acquired by Taiwan.

F-86E/F/Mitsubishi F-86F-40

NAA began work on the F-86E (NA-170) on 15 November 1949. It differed from the late-production F-86A only in having fully power-operated controls for improved manoeuvrability at high speeds. The most innovative feature of this system was the so-called 'all flying' tailplane, or controllable horizontal tail. Though the rudder was cable-controlled, the elevators and horizontal stabiliser were controlled and operated as one unit by hydraulic power in response to control column movements. (On the F-86A the stabiliser was mechanically adjustable for trim control.) The new system eliminated many of the undesirable effects of compressibility such as loss of control at high Mach numbers. On 17 January 1950, NAA received a USAF contract for 111 F-86E aircraft. The first F-86E-1 (50-579) was flown by George Welch on 23 September 1950 and was accepted by the USAF on 9 February 1951.

Capt. Manuel J. 'Pete' Fernandez of the 334th FIS, 4th FIW was patrolling the skies over Korea in F-86F 51-2857 during a routine training mission on 12 February 1954. *(via Larry Davis)*

Acceptance of 60 F-86E-l Sabres began that same month and was followed by 51 F-86E-5 models, which differed from the -1 only in having minor changes to the cockpit instrumentation layout. The first F-86Es were issued to ADC in April 1951 when they equipped the 97th FIS at Wright-Patterson AFB and the 23rd FIW at Presque Isle AFB. Others equipped the 60th FIS, 33rd FIW at Otis AFB, Massachusetts, and in June and July that year shipments commenced to the 4th FIW in Korea, where the aircraft eventually replaced this unit's F-86A Sabres. On 1 and 9 November 1951, 75 more F-86Es were hurriedly shipped to Korea to replace the F-80s of the 51st FIW.

The first production F-86E-1-NA was accepted for the USAF by Major Charles E. Yeager in March 1951 and 396 aircraft had been built by April 1952. The F-86E was powered by the 23.12-kN (5,200-lb) thrust J47-GE-13 and from the beginning, NAA had planned to supplant this engine with the more powerful 26.28-kN (5,910-lb) thrust J47-GE-27 in an improved version designated F-86F (NA-172). Work on the new aircraft began on 31 July 1950 and it was scheduled for production as the F-86F in October 1950. No mock-up was constructed, since the airframe remained the same, although an entirely new engine installation was required.

Above: F-86F-30 52-5092 is shown at Williams AFB in the 1950s. *(via Larry Davis)*

Below: F-86F 52-4608 was used as a test bed for the Rocketdyne AR2-3 motor. *(via Larry Davis)*

An original contract for 109 aircraft was increased to 360 by 30 June 1951 and it was planned to manufacture F-86Fs at Columbus, Ohio. However, General Electric experienced problems with the powerplant and so NAA installed the J47-GE-13 in the first 132 NA-172s, which, beginning in August 1951, were delivered to the USAF as the F-86E-10-NA. Outwardly, F-86E-10s were distinguishable by a flat windshield in place of the vee-shaped windscreen fitted to the F-86A and other E models.

The first F-86F-1 (51-2850) was flown on 19 March 1952 and by late that summer the first of 78 examples was in service with the 84th Squadron at Hamilton Field, and the 51st and 4th FIWs in Korea. The F-86F-l was followed by sixteen F-86F-5 machines fitted with underwing shackles to carry 200-US gal (757-litre) tanks in place of the earlier 120-US gal (454-litre) tanks. These increased the combat radius from 531 to 745 km (330 to 463 miles). A change of gunsight, from the Mk 18 gyro sight to the A-4, characterised the F-86F-10 production block. In use in Korea, the Mk 18 sight required the pilot to manually operate the ranging control which was not ideal for high-speed deflection firing. The A-1CM, which had first appeared on late F-86A models and which was fitted to all F-86F models, had radar for automatic ranging, but was often unserviceable because of poor maintenance. The A-4 operated in much the same way and with the same radar as the A-1CM, but was simpler to maintain in the field. The last 100 aircraft on the NA-172 contract were F-86F-15s with re-positioned control systems, but in April 1952 more delays with the J47-GE-27 led to every aircraft in this block, from the 8th production aircraft onwards, being fitted with the older -13 engine. Deliveries to the USAF began in August 1952, where the aircraft equipped only training units because by now, J47-GE-27 powered F-86Fs were entering first-line service. The 93 F-86E-15s completed by December 1952 brought the F-86E total to 396 aircraft. Meanwhile, a contract for 441 F-86F (NA-176) production aircraft had been approved

The first production FJ-3 was first flown on 11 December 1953, but by July 1954 only 24 FJ-3s had been accepted. These were used to equip VX-3 and VF-173. On 8 May 1955, VF-173 landed its FJ-3s aboard the carrier USS *Bennington* in the Atlantic. *(via Larry Davis)*

on 17 March 1951, but the first aircraft (51-13070) did not fly until May 1952 and delivery of the 100 F-86F-20s was not completed until January 1953. The F-86F-20 could carry two 200-US gal drop tanks, had armour protection around the horizontal stabiliser control system, and a different radio and cockpit arrangement than on previous Sabres.

The next version to appear, on 26 October 1951, was the NA-191 fighter-bomber or 'dual store' version, which had provision for two drop tanks under each wing. A 120-US gal tank or a bomb of up to 454 kg (1,000 lb) could be carried on each of the inner underwing strongpoints, while the outer pair could carry 200-US gal tanks. A USAF contract was approved on 5 August 1952, for 907 aircraft to be built at Inglewood and the new configuration was also to be incorporated on 341 NA-176 aircraft already on order at Columbus and a further 259 NA-193 aircraft ordered on 17 October 1952. Deliveries of the first F-86F-30s were begun at Inglewood in October 1952 and by January 1953 the Columbus-built F-86F-25 version had appeared.

Meanwhile, NAA project engineers, led by Fred R. Prill, explored ways of improving F-86 performance to allow pilots to make tighter turns at high altitudes during combat with the MiG-15 in Korea. At the suggestion of test pilots they tried a fixed leading edge, eliminating the

USMC FJ-3M Furys 136134 and 139211 were photographed in flight. *(via Larry Davis)*

drag-producing effects of the slats. In August 1952 three Sabres were tested with non-slatted leading edges, extended 6 in (15.24 cm) at the root and 3 in (7.62 cm) at the tip and popularly referred to as the '6-3' extension or '6-3' wing. Changes of airflow across the wing required the addition of 12.70-cm (5-in) high wing fences at 70 per cent of the span and wing area increased from 26.75 m² to 28.08 m² (287.90 to 302.30 sq ft), but the improvement in performance was immediate. Lower drag from the smoother wing entry increased the F-86F's top speed from 1107 to 1118 km/h (688 to 695 mph) at sea level and from 972 to 978 km/h (604 to 608 mph) at 10668 m (35,000 ft), while range was also slightly increased. Most importantly of all, manoeuvrability at high altitudes and high Mach numbers improved significantly. By delaying buffet, the new wing gave the pilot an increase in usable *g*, allowing him to fly closer to the maximum *g* limit before buffet commenced. The down side of this was the loss of low-speed characteristics associated with the slatted wing. Stall speed on take-off rose sharply from 206 to 232 km/h (128 to 144 mph) and a yaw-and-roll effect at low speeds preceded the stall. A final landing approach speed of about 278 km/h

(173 mph), with 222 km/h (138 mph) at touchdown had to be flown and this required a longer landing roll. In September 1952, 50 kits were shipped to Korea to convert F-86Fs to the new '6-3' configuration and enough kits were ordered to convert all the F-86Fs in Korea. All production aircraft, starting with the 171st F-86F-25 (51-13341) and the 200th F-86F-30 (52-4505) were also fitted with the extended leading edge. The extended-wing version made a huge difference to the air fighting in Korea and allowed the F-86 Sabre pilots to enjoy an even greater superiority over the MiG-15.

Altogether, 135 F-86F-1 to -15 'single-store' Sabres at Inglewood and 100 F-86F-20s at Columbus, were built. These were followed in the 1953 to early 1954 period by 600 Columbus-built F-86F-25s and 859 F-86F-30s built at Inglewood with the 'dual-store' provision. By July 1953 there were 13 Sabre day fighter wings active with the USAF, including four operating in Korea.

FJ-2 Fury

The Korean War revealed that neither the Grumman Panther nor the McDonnell Banshee could operate in MiG-dominated airspace without unacceptable losses and so the USN showed renewed interest in carrierborne swept-wing fighters. One of these was the NA-181, an NAA design proposal for a navalised version of the F-86E-10 Sabre. On 10 February 1951 the Bureau of Aeronautics issued a contract for two XFJ-2 and one NA-185/XFJ-2B aircraft. The first two prototypes, BuNos 133754 and 133755, were F-86E airframes with their armament deleted, and modified for carrier operations with a catapult spool in the belly and an 'A'-frame type arrester hook at the rear fuselage. They were also fitted with lengthened nose oleos to compensate for the high angle of attack required for a catapult launch. XFJ-2B BuNo. 133756 was used purely as an armament test airframe at the Ordnance Test Station, Inyokern, to test a four Colt Mk 12 20-mm cannon installation and was not navalised. NAA also received a simultaneous production order for

300 FJ-2 Furys, which would be built at its new Columbus, Ohio, plant where, in 1952, F-86F production was in full swing. The XFJ-2B was the first to fly when test pilot Robert Hoover took it aloft on 27 December 1951. In early 1952, the two XFJ-2 aircraft were delivered to the USN Test Center at Patuxent River. Evaluation trials were carried out aboard USS *Midway* in August 1952 and carrier qualification trials (carquals) were completed on the USS *Coral Sea* in December 1952.

The first production FJ-2 Fury (the USN equivalent of the F-86F) was accepted by the Navy in October 1952. Power was provided by the 26.68-kN (6,000-lb st) J47-GE-2 navalised version of the J47-GE-27. Additional naval equipment raised the take-off weight to 8524 kg (18,791 lb) compared to the F-86F's 8073 kg (17,797 lb). Maximum speed was 1088 km/h (676 mph) at sea level, and 969 km/h (602 mph) at 10973 m (36,000 ft). A Mk 16 Model 2 sight and AN/APG-30 radar aimed the four 20-mm cannon, which were supplied with 150 rounds per gun. When the war in Korea was nearing its end, the original order for 300 Furys was eventually reduced to one for 200 aircraft. Since the Navy preferred the F9F-8 Cougar, which had a better deck performance, the entire FJ-2 production run was given over to the US Marine Corps (USMC). By the end of 1952, only five aircraft had been completed at Columbus and it was not until January 1954 that the first FJ-2 was delivered to the USMC when 25 examples went to VMF-122 at Cherry Point, North Carolina. Production continued until September 1954 and by 1955 the FJ-2 equipped six USMC squadrons. In June 1954 VMF-235's FJ-2s carried out operational trials aboard USS *Hancock*, the first US carrier to be fitted with the British-developed steam catapult. The FJ-2 had a short operational career with the USMC. By 1956 the type had been replaced in front-line service and the FJ-2s in Navy Reserve units finally went in 1957.

FJ-3 Fury

By March 1952, NAA was only too well aware of the FJ-2's unsuitability for carrier operations and work began on the FJ-3. This was powered by a 34.24-kN (7,700-lb) thrust Wright J65-W-2 engine (a licence-built version of the Armstrong Siddeley Sapphire), which was tested in the fifth production FJ-2. The FJ-3 also had a larger air intake than the FJ-2 and ammunition for the four 20-mm guns was increased by 48 rounds. The first production FJ-3 was completed at Columbus on 11 December 1953 and flown that same day by test pilot Bill Pearce. A production order covering 389 FJ-3s (135774/136162) for the USN was received, but, again, delivery was slow. By July 1954 only 24 FJ-3s had been accepted. These were used to equip VX-3 and VF-173, which carried out the Fleet Introduction Program at Patuxent River in a record time of just 29 days. On 8 May 1955, VF-173 landed its FJ-3s aboard the carrier USS *Bennington* in the Atlantic. On 22 August that same year VX-3 began operational evaluation of the mirror landing system installed aboard *Bennington*, the first landing using the system being made by the VX-3 CO, Cdr R. G. Dose.

In 1955, the FJ-3 Fury underwent a number of modifications to try to cure its poor carrier performance. The wing slats were replaced by extended leading edges, which housed an extra

469 litres (124 US gal) of fuel, and the number of underwing hardpoints was increased from two to six, enabling the FJ-3 to carry 500- or 1,000-lb (227- or 454-kg) bombs, rocket packs or additional fuel tanks. Sidewinder air-to-air missiles (AAMs), first tested in 1952, were fitted to the Fury (as the FJ-3M) from the 345th aircraft onwards. By August 1956, the FJ-3/3M Fury formed the equipment of 18 USN and four USMC squadrons.

FJ-4 Fury

The design of the FJ-4 long-range attack fighter began at the Columbus plant in February 1953 and in June that year an order for two prototypes, under the company designation NA-208, was received. In order to satisfy the need for extended range the Fury's fuel capacity was increased by 50 per cent, but this entailed a major redesign of the entire airframe to compensate for the resulting substantial increase in gross weight. The wing and tail surfaces were made thinner and their span and area were increased. Extra armour was installed in the

FJ-4B (NA-244) Fury 143494 served with VX-4. *(via Larry Davis)*

nose, using space made by reducing the amount of 20-mm ammunition carried. A new levered-suspension undercarriage, increasing the track to lift 17.78 cm (7 in), was installed and the fuselage and cockpit contours were revised. The fuselage was deeper than that of the FJ-3 and included a dorsal spine. On 16 October 1953, NAA received a production order for 150 FJ-4 (NA-209) aircraft. The first of the two FJ-4 prototypes (139279 and 138280) was flown by test pilot Richard Wenzell on 28 October 1954, powered by a Wright J65-W-4. Production aircraft were fitted with the 34.24-kN (7,700-lb st) J65-W-16A that gave a maximum speed of 1094 km/h (680 mph) at sea level and 1014 km/h (630 mph) at 10973 m (36,000 ft). The first FJ-4s were delivered to VMF-323 in September 1956 and by the end of March 1957 all the FJ-4s had been delivered, to seven more USMC fighter and attack squadrons.

FJ-4B Fury

The FJ-4B made its first flight On 4 December 1956. It differed from the previous model in having a strengthened wing to accommodate the increase from four to six underwing stores positions. The new aircraft was also equipped with a Low Altitude Bombing System (LABS) for the delivery of a tactical atomic weapon. Additional speed brakes were also fitted under the rear fuselage, to provide better control at low level. The Fleet Introduction Program for the FJ-4B was undertaken by VA-126 and VMA-223, with particular emphasis on low-level nuclear weapons delivery. The FJ-4B had an inflight-refuelling probe installed in its port wing, and from June 1957 underwing fuel packs came into use under the so-called 'buddy-buddy' in-flight refuelling system, enabling a Fury to take on an additional 1435 kg (3,163 lb) of fuel from an aircraft of the same type. The FJ-4B equipped three Marine Corps' squadrons and served with ten USN Pacific Fleet squadrons. In October 1958 24 FJ-4s of VMA-212 and VMA-214 completed Operation *Cannonball*, the first Trans-Pacific crossing by single-seat naval aircraft, flying from MCAS Kaneohe to NAS Atsugi, Japan, with

stopovers at Midway and Guam, and inflight refuelling from Boeing KB-50 tankers off Wake Island, and NAA AJ-1 Savage aircraft near Iwo Jima. On 25 April 1959 FJ-4Bs of VA-212 became the first to deploy overseas with the Martin Bullpup precision air-to-surface missile, sailing from Alameda on board USS *Lexington* to join the Seventh Fleet in the western Pacific. The Fury could carry five Bullpups, plus the associated equipment pack. VA-126 was the last first-line squadron to operate the Fury, in September 1962, and some continued with Reserve units until 1964. All told, 1,115 Furys were delivered between January 1952 and May 1958.

F-86D 'Sabre Dog'

The F-86D (NA-164) differed radically from earlier models in having a 0.76-m (30-in) fibreglass radome housing the antenna of the Hughes AN/APG-36 (AN/APG-37 in later models) search radar, above the nose intake. In May 1948 NAA envisaged that the F-86D design would be a two-seat interceptor armed solely with missiles, with the second crew member assisting the pilot in making radar-controlled interceptions and being responsible for navigation. However, this was before improved automated AI radar, which reduced the operator's workload, became available. Aircraft performance limitations would also result if a second crew position were included, and so this and other problems, such as restricted fuselage fuel tank space, led to the two-seat concept being abandoned. This placed greater reliance on automated systems such as the E.3 and later E.4 fire control systems, under development by the Hughes Aircraft Corporation, being made available for single-pilot operation to work efficiently. A complex engine control system also had to be developed for operation of the afterburning J47-GE-17 engine when it was realised early in the aircraft's design that pilots would have difficulty achieving the correct engine performance and particularly the use of afterburner, during say, AI radar interceptions. NAA and General Electric jointly began the

development of a single throttle lever control, fitted with an electronic fuel selector which could determine the amount of fuel to the engine and afterburner to maintain maximum efficiency even when the throttle was slammed open and closed rapidly.

On 28 March 1949, engineering design work on the F-86D began. The wing, though generally similar to that of the F-86A, was strengthened. The fuselage was larger, the vertical tail surfaces were increased in area and a slab-type tailplane was adopted for improved longitudinal control. Once the USAF had shown an immediate interest in the project, work on the NA-165 production version started on 7 April and on 1 June NAA began building a mock-up. News that the Soviet Union had detonated a nuclear device gave the all-weather interceptor programme an added sense of urgency and on 7 October NAA received a letter of intent for 122 production F-86Ds. (A formal contract was approved on 2 June 1950, when the total on order was increased to 153 examples.) On 28 November 1949, the first of two YF-86D prototypes (50-577/578) went to Muroc (later Edwards AFB). The first flew on 22 December with George Welch at the controls and all appeared well, but the test programme was delayed when the aircraft was damaged after an

F-86D-40 52-3863 *Dennis The Menace*, of the 97th FIS, was seen at Wright-Patterson AFB in 1954, wearing the unit's stylised 'Devil Cat' red fuselage flash. *(NAA)*

undercarriage problem. It did not resume until 17 October 1950.

Problems with the first production E-3 and E-4 fire control systems, radar, and rockets (when the pilot selected a ripple-firing sequence), engine controls and autopilots, caused further delays. Flight-testing revealed aerodynamic problems created by changes of design from the original Sabre. All the structural changes had increased drag and a number of aerodynamic refinements had to be incorporated in production aircraft. To reduce drag, vortex generators were positioned around the fuselage and tail assembly to create a vortex pattern downstream of their location, re-energising the slow-moving boundary layer airflow and delaying its separation from the aircraft's surface.

F-86D pilot workload was high and the aircraft required more pilot training than any other type in USAF service. An F-86D all-weather school was established at Perrin AFB, Texas, which was equipped with an Erco flight simulator that duplicated all the functions of the F-86D cockpit, including attack modes. Live firing of the 2.75-in (70-mm) rockets was made at a banner target, which was a 9- x 1.8-m (30- x 6-ft) plastic mesh panel fitted with metal discs to produce echoes on the F-86D's AI radar, and was normally towed behind a B-45 Tornado.

The F-86D finally entered service with active ADC units in April 1953 and by the end of 1953 600 F-86Ds were in service. Some 2,504 F-86Ds were built for the USAF and by mid-1955 1,026 of ADC's 1,405 interceptors were F-86Ds, equipping 20 ADC wings. Late in 1953 some Sabre Dogs were assigned to the 5th AF in Korea but being much heavier than earlier F-86s, they had difficulty operating from the rudimentary South Korean airstrips and were soon withdrawn. Production F-86Ds were fitted with a 4.75-m (15-ft 7¼-in) diameter ribbon drag 'chute, beginning with the F-86D-45 that entered service in April 1954. This reduced the aircraft's landing roll from 777 to 488 m (2,550 to 1,600 ft), considerably enhancing the safety factor in all runway conditions.

In the continental United States, F-86Ds equipped 19 Fighter (All-Weather) Wings and their component squadrons. After first-line service, the F-86Ds were also allocated to ANG units. F-86Ds also served in the UK, equipping the 406th FW at RAF Manston from November 1953. The 512th FIS moved to Soesterberg, Netherlands, in November 1954, but the rest of the 406th remained at Manston until May 1958, when the Wing was deactivated. By this time, all-weather defence of UK airspace had been taken over by the Gloster Javelin. On 5 July and 1 September 1958, respectively, the 497th and 431st FISs at Torrejón and Zaragoza in Spain, were transferred from USAFE control to SAC control, in order to provide air defence for B-47s deployed to Spanish bases on *Reflex Alert* missions. Each squadron had an establishment of 24 Sabres. On 1 July 1960, the units reverted back to USAFE control and soon re-equipped with the Convair F-102A Delta Dagger.

F-86G

This designation was applied originally to the F-86D-20-NA when it was proposed to install the J47-GE-33 engine in this aircraft.

F-86H

On 3 November 1952 NAA received a contract for 175 (NA-187) F-86H-1 aircraft, a more powerful fighter-bomber version of the F-86F.

While F-86H-10 (NA-203) 53-1298 is illustrated, the F-86H-5 had appeared in January 1955 with four 20-mm M-39 guns. Developed by the USAF, NAA and the Ford Motor Company, this weapon had a revolving-drum feed that discharged the F-86H's ammunition load (600 rounds) in six seconds. Sixty F-86H-5s were followed at Columbus by the 300 F-86H-10s, which differed only in having electrical system modifications. *(NAA)*

Two pre-production models and a static test airframe were to be built at Inglewood and the remainder at Columbus. Intended primarily for low-level attack, the NA-187 differed from previous Sabre models in having the larger and more powerful 39.67-kN (8,920-lb st) J73-GE-3/3A turbojet. The basic Sabre airframe had to be extensively strengthened to accommodate the new engine. As the air intake duct sizes had already reached an upper limit for the size of the airframe, NAA enlarged the air intake and split the fuselage lengthwise along a theoretical waterline, splicing in an extra 0.15 m (6 in) of fuselage depth, just as CAC in Australia had done when installing the Rolls-Royce Avon in the CA-27 Sabre. Other notable external differences included a clamshell-type cockpit canopy; a larger, power-boosted tailplane without dihedral; a heavier undercarriage and improved suspension and release mechanism for underwing ordnance loads. Fuselage fuel tanks were fitted to supplement the normal wing tanks so that although fuel consumption was higher than in previous models, the combat radius remained much the same at around 1014 km (630 miles).

F-86H-1 52-1975, the first of two prototypes (the other was 52-1976) was flown for the first time by NAA test pilot Joseph Lynch at Edwards AFB on 30 April 1953. Originally it had the '6-3' extended leading edge and wing fences, but by December 1953 the slats had been deleted. The first Columbus-built F-86H-1 (52-1977) had extended leading edges and was flown on 4 September 1953, but production of the new model did not get underway until after May 1954 when the last of 700 F-86Fs were completed. The F-86H-1 was armed with six 0.5-in (12.7-mm) machine-guns as in previous Sabre models, but the F-86H-5 and subsequent production batches were fitted with four 20-mm M-39 cannon.

On 11 June 1953, NAA received contracts for a further 300 F-86H-10 (NA-203) aircraft to be built at Columbus. Like the last 264 of 1,259 F-86Fs (F-86F-35), the F-86H was equipped to carry a 1,200-lb (544-kg) atomic bomb under the port wing – the drop tanks being carried under the starboard wing. To avoid destruction of the aircraft at the time of release a special delivery technique called 'toss bombing' was to be used, the exact moment to release the bomb accurately being automatically computed by a LABS (Low Altitude Bombing System) installed in all F-86H Sabres beginning with the fifth aircraft. Controls were provided for arming and disarming the device in flight. Conventional weapons that could be carried on the F-86H included two 1,000-lb (454-kg) or smaller bombs, two 750-lb (340-kg) napalm bombs, or eight 5-in (127-mm) HVARs. Radar ranging (AN/APG-30) for the A-4 gun and rocket sight and P-2 strike camera, was standard.

Ten F-86H-ls had been accepted by the end of June 1954, but operational testing was delayed by accidents. On 24 May 52-1982 the sixth production aircraft, was lost at Edwards AFB. Deliveries resumed on 2 August, but on 25 August the leading Korean ace, Capt. Joseph McConnell, was killed at Edwards while flying 52-1981. McConnell experienced a total hydraulic failure and tried to recover to base using just the throttle and rudder. He almost made it, but turbulence on his approach lifted one wing and he tried to eject, but his 'chute was unable to deploy fully at such a low altitude and he was killed. On 1 October operational testing finally took place at the Air Proving Ground at Eglin AFB, Florida, where the F-86H's shorter take-off run, higher rate of climb and greater ceiling, wider combat radius, and better air-to-ground gunnery characteristics revealed it to be a better fighter-bomber than the F-86F. Increased power provided faster acceleration and cruising speed, but because of the airframe's Mach limitations, added little to the maximum speed except above 10668 m (35,000 ft). The F-86H-1 entered operational service with the 312th FBW at Clovis AFB, New Mexico in November 1954. Altogether, six USAF wings were equipped with the F-86H, the last being the 4th Fighter (Day) Wing, which reformed at Seymour Johnson AFB in December

1957. The F-86H also equipped no less than 17 ANG units and for a short time in 1957, large numbers of F-86Hs were issued to the Air Force Reserve (AFRES) until it was decided to wholly re-equip the Reserve for a transport role.

USAF pilots reported that as an air-to-air fighter, the F-86H was unequal to the F-86F at all altitudes because its higher wing loading made it incapable of sustaining the *g* forces needed for tight turns at high altitudes. Turning was improved with the installation, on the 15th F-86H-1, of the '6-3' leading edge, extended wingtips and wing fences. Wing span was increased from 11.28 m (37 ft) to 11.92 m (39 ft 1¼ in) and wing area to 29.11 m² (313.40 sq ft). The last F-86H-10 (53-1528) was delivered to the USAF on 16 March 1956, the last ten aircraft having slats on their extended leading edges for greatly improved low-speed handling. The F-86H enjoyed only a relatively brief career in the USAF. The 413th, 312th and 474th FDWs converted to the F-100 Super Sabre in 1956 and the 50th FBW, which was based in France, received F-100 aircraft in 1957. The 4th FDW continued to operate the F-86H only until early 1958, when it too received F-100s.

CL-13 Sabre Mk 1 (F-86A-5)

A single aircraft (19101) powered by a 22.24-kN (5,000-lb st) J47-GE-13 was assembled by Canadair from US-made components.

CL-13 Sabre Mk 2 (F-86E-6)

This was the first Canadian production version and the first of these aircraft (RCAF 19102) was flown from Cartierville early in March 1951. Some 350 Sabre 2s were built. All except 60 (52-2833/2892), which were sent to the USAF to equip the 4th and 51st FIWs in Korea, and three, which went to the RAF, were delivered to the RCAF. Essentially, the Mk 2 was the same as the Sabre 1, but was fitted with the all-flying tailplane and A-1C gunsight. Armament (which remained the same on all Canadian-made versions) was six 0.5-in (12.7-mm) M-3 machine-guns, two 1,000-lb (454-kg) bombs or equivalent in other bombs or military stores or 16 5-in (127-mm) HVARs. External tanks could be fitted to give a ferry range of 1,541 nm (2856 km; 1,774 miles).

CL-13 Sabre Mk 3 (F-86J)

One Sabre 2 (RCAF 19200) was modified to take a 26.68-kN (6,000-lb st) Avro Orenda 3. It had no armament but was otherwise identical to the Sabre 2.

CL-13 Sabre Mk 4

This was essentially identical to the Sabre 2 but

F-86H 52-2068, of the 131st TFS, Massachusetts ANG, departs Barnes Field for Phalsbourg AB, in 1961. *(via Jerry Scutts)*

F-86H-10 0-31503 (53-1503), of the 104th TFS, Maryland ANG, was photographed at Glenn L. Martin State Airport, Baltimore. The 104th TFS received its first Cessna A-37B to commence conversion from the F-86H on 25 April 1970. *(APN via Jerry Scutts)*

with some equipment changes, including improved air conditioning, and was delivered to both the RAF and RCAF.

CL-13A Sabre Mk 5

This Sabre variant was powered by a 29.35-kN (6,600-lb st) Avro Orenda 10 and was fitted with an extended leading edge (with no slats), wing fences, a vee-shaped windshield, an A-4 gunsight and an improved oxygen system. One Sabre 5 was modified with an area-rule fuselage and first flew on 13 July 1954.

CL-13C Sabre

This modified Sabre Mk 5 (RCAF 23021) was built following the disclosure of Richard T. Whitcomb's area rule, which indicated that the speed of an aircraft could be increased by adding volume to the fuselage. At the request of the National Research Council (NRC) aluminium blisters were fitted to the aircraft's fuselage and after initial tests by Canadair, were tested by the National Aeronautical Establishment (NAE) at Uplands Airport, Ottawa. The performance was compared with a standard Sabre 5 (RCAF 23275) and found to be identical, but this was achieved with a fuselage of about 10 per cent greater volume.

CL-13B Sabre 6

This Sabre version was powered by a 32.35-kN (7,275-lb st) Avro Orenda 14 and was identical in all other respects to the Sabre 5 except for having slats (although some early Mk 6s were built without slats). A single CL-13E Sabre 6 (RCAF 23544) was fitted with the NRC reheat system, installed at the request of NRC to test its system in which a portion of the fuel was injected upstream of the engine turbine. When tested by the NAE at Uplands, the system developed an additional 215 per cent gross thrust at all altitudes.

The CL-13D Sabre 6 with an Armstrong Siddeley rocket motor added for improved climb and acceleration; the CL-13G two-seat trainer; the CL-131-1 with all-weather radar system; and the CL-13K/Sabre 7 with an Avro Orenda 14R engine incorporating a reheat installation, were never built.

F-86J

This designation was briefly applied to the Canadair-built Sabre Mk 6 by the USAF.

F-86K

Late in 1952, Air Materiel Command asked NAA to investigate the possibility of redesigning the F-86D to produce a simplified all-weather two-seat fighter that could be built in Europe and issued to NATO air forces. NAA decided that a two-seat arrangement was not feasible and so retained the single-seat layout of the F-86D. As a single-seat aircraft the F-86K would prove far from ideal as an all-weather interceptor, but it would be instrumental in helping European NATO air forces make the transition from transonic to supersonic combat aircraft. On 14 May 1953, work on the F-86K (NA-205) began at Inglewood. The F-86K, which was powered by a J47-GE-17B turbojet, differed from the F-86D in two main respects. The NAA MG-4, a more simplified, yet extremely effective system, that used the same AN/APG-37 radar antenna, replaced the E-4 fire control system. (The MG-4 computed a course for lead-pursuit attack and automatically provided the pilot with firing range and break-off time.) Also, four 20-mm

M-24A-1 cannon, with 132 rounds per gun, replaced the D's Mighty Mouse missile tray. The movement of the armament forward required an 0.20-m (8-in) extension of the fuselage to preserve the centre of gravity location. (Towards the end of their career, some F-86Ks received the addition of two Sidewinder AAMs.)

On 16 May 1953, a contract for the licence manufacture of the F-86K by Società per Azioni Fiat of Turin was signed and on 28 June 1954 another agreement enabled the assembly of an initial batch of 50 aircraft from components built at Inglewood, with funds supplied by MDAP. In December 1953, a contract was placed with NAA for the assembly of 120 F-86Ks for delivery to the Netherlands and Norway. (By 1955, 59 F-86Ks had been delivered to the Netherlands, where they equipped Nos 700, 701 and 702 Squadrons of the KLu; 60 went to Norway, where they equipped Nos 334, 337 and 339 Squadrons of the RoNAF and one (54-1231) was retained by NAA for trials work.)

Two YF-86K-1 prototypes (52-3630 and 52-3804) were converted on the assembly line from standard F-86D-20-NA airframes, and on 15 July 1954 52-3630 was flown for first time at Los Angeles by NAA test pilot Raymond Morris. The two prototypes were subsequently sent to Italy as pattern aircraft and were later operated by the AMI (Italian air force). The first Fiat-built F-86K flew on 23 May 1955. In all, 221 F-86Ks were assembled by the Italian company, the last batch of 45 (NA-242) aircraft having their

This aircraft is an F-86K of the Aeronautica Militare Italiana. *(AMI via Jerry Scutts)*

CL-13 Sabre 6 XB920/K (ex-XB944) formates with a Hunter F.Mk 4, both aircraft being from No. 112 Squadron, 2nd ATAF. The Sabre is painted with the shark-mouth markings that the Squadron used on its Kittyhawks in the Western Desert in World War Two. No. 112 Sqn equipped with the Mk 4 Sabre at Brüggen early in 1954 and began receiving its first Hunters in April 1956. The Hunter cockpit was smaller and more cramped than that of the Sabre, but pilots were pleased to have the Martin Baker ejection seat with a face blind, an improvement, it was considered, on the American 'bang seat' which was operated from the arm rests. By the end of June 1956, Hunters had replaced all RAF Sabres in Germany. (via Brian Pymm)

wingtips extended by 0.30 m (1 ft) and being fitted with leading-edge slats. These modifications increased the all-up weight by 386 kg (850 lb), but the slats and increased wing area improved overall stability and manoeuvrability dramatically. An earlier tendency to roll and yaw at low speeds was eliminated; the stalling speed was reduced from 232 to 200 km/h (144 to 124 mph) and the take-off run was reduced by 244 m (800 ft). All 59 F-86Ks operated by the KLu were retrofitted with slats and wingtip extensions. Some 63 of the F-86Ks assembled by Fiat were delivered to the AMI where they equipped the 21°, 22° and 23° Gruppi of the 51 Aerobrigata Caccia Ogni Tempo (All-Weather Fighter Air Brigade). Fiat exported 88 F-86K Sabres to the Federal German Republic, where they equipped the Luftwaffe's Jagdgeschwader 74 at Neuburg, and 60 were delivered to the Armée de l'Air (French air force), equipping 13 Escadre. Six F-86Ks were delivered to the KLu and four went to the RoNAF as attrition replacement aircraft.

F-86L

Early in 1956, F-86Ds were progressively withdrawn from ADC and upgraded to F-86L standard by the fitting of updated equipment such as the AN/ARR-39 datalink receiver, an AN/ARC-34 command radio, an AN/APX-25 IFF transponder and AN/ARN-31 glide slope receiver. F-86D-10 and F-86D-40 aircraft so modified were redesignated F-86L-11 and F-86L-41 respectively, and F-86D-43 and F-86D-60 machines became F-86L-45 and F-86L-60 aircraft, respectively. Outwardly, the F-86L differed little from the F-86D, except that span was increased to 11.92 m (39 ft 1¼ in) with the addition of 0.30-m (12-in) wingtip extensions, which gave a much improved turning

performance at altitude. The first F-86L flew in October 1956 and the 'production' run was used to equip ADC units until the F-102A and F-106A interceptors became available. From 1957, the all-weather Sabres were assigned to 23 ANG squadrons, the last being the 196th Ontario, California FIS, in 1965.

TF-86F

The TF-86F (NA-204) two-seat transonic trainer developed from the basic F-86F-30, was a proposal put forward by NAA early in 1953 to bridge the gap between the subsonic Lockheed T-33A and the single-seat Sabre fighter. The second cockpit was accommodated by lengthening the fuselage by 1.60 m (5 ft 3 in) and moving the wing forward 0.20 m (8 in). The first of two TF-86Fs (F-86F-30 52-5016), both of which had a long, clamshell-type canopy, was flown on 14 December 1953; it was destroyed later when the pilot failed to recover from a roll during a demonstration flight. The second TF-86F (53-1228), which had a dorsal fin to increase airflow over the rudder for improved handling at low speed and high angles of attack and was armed with two 0.5-in (12.7-mm) machine-guns for gunnery practice, was flown on 17 August 1954. The TF-86F did not enter production.

CA-26 Sabre

This designation was given to the single prototype ordered by the Australian Government from Commonwealth Aircraft Corporation (CAC), at Melbourne, during 1951 on Contract CA-26 (A94-101). It was constructed mainly of imported components, powered by a Rolls-Royce RA.7 Avon engine and armed with a pair of 30-mm

The second TF-86F (53-1228), which had a dorsal fin to increase airflow over the rudder for improved handling at low speed and high angles of attack, and was armed with two 0.5-in (12.7-mm) machine-guns for gunnery practice, was flown on 17 August 1954. The TF-86F did not enter production. *(via Larry Davis)*

ADEN guns. It first flew on 3 August 1953 and was used for type and flight trials with CAC and the RAAF during the period 1953 to 1955, before being relegated to airframe instructional duties.

CAC CA-27 Sabre Mk 30

These aircraft represented the first production batch of 21 aircraft (A94-901/921) ordered on CA-27 Contract and built during 1953 to 1954. A94-901 first flew on 13 July 1954, powered by a Rolls-Royce Avon engine. The remaining aircraft were powered by Rolls-Royce Avon 20 engines using imported components and were built with leading-edge wing slats.

CAC CA-27 Sabre Mk 31

This designation applied to those aircraft in the CA-27 production batch which, in 1957 to 1958, were converted to Mk 31 standard by the removal of their wing slats and the addition of '6-3' extensions.

A production batch of 21 new-build aircraft (A94-922/942) was also built. These had the Avon 20 engine and standard '6-3' wing. A94-938 and A94-942 were modified with 318 litres (70 Imp gal) of fuel in their leading edges.

CAC CA-27 Sabre Mk 32

These aircraft consisted of a production batch of 69 machines A94-943/990 and A94-351/371, powered by the Australian-built Rolls-Royce Avon 26 engine and fitted with 272-litre (60-Imp gal) 'wet leading edges' and dual store capability. A twin Sidewinder missile installation was retrofitted to all Mk 32 aircraft.

QF-86E

This designation applied to 56 remotely piloted ex-RCAF CL-13 Sabres for use as high-speed target drones.

5. Overseas Sabres

Argentina

Beginning in May 1960 the *Fuerza Aérea Argentina* (FAA, or Argentinian air force) received 28 F-86F-30 Sabres, refurbished by NAA, to replace outdated Gloster Meteor F.Mk 8 aircraft which had been its main interceptor/strike aircraft since 1948. The F-86Fs were assigned to Grupo 6 de Caza (6th Fighter Group). Seventeen Sabres were lost in FAA service, the aircraft remaining in service until 19 June 1986.

Australia

In October 1951 the Australian government obtained a manufacturing licence for the F-86F for the RAAF. However, the Australians considered the F-86F underpowered and its armament lacked punch, so the J47-GE-27 engine was replaced with a 33.35-kN (7,500-lb) thrust RA.7 Avon turbojet and the six 0.5-in (12.7-mm) machine-guns were replaced with two 30-mm ADEN cannon. The decision to use the Avon powerplant caused some airframe design changes. Being broader and shorter than the American powerplant and weighing 181 kg (400 lb) less, the British engine would alter the Sabre's centre of gravity position unless it was repositioned further aft in the fuselage. This meant that to fully support the repositioned engine and its jet pipe, the rear fuselage had to be shortened by 0.66 m (2 ft 2 in) and the forward fuselage increased by the same amount. To achieve this the rear fuselage had to be completely redesigned, to avoid the tailplane's inertial loads and tail unit flight loads being

transferred to the powerplant. (Later, a minor centre of gravity problem was solved by weighting the nose of the prototype with 36 kg (80 lb) of lead ballast, while in the production model a pump supplying emergency power to the controls was moved further forward.) The Avon's greater power also meant that the air intake had to be increased in size. This was achieved by splitting the front fuselage horizontally and lowering the bottom line 0.08 m (3 in) to avoid having to make extensive changes to the cockpit area. All these modifications meant that only about 40 per cent of the original F-86F fuselage structure remained.

The Commonwealth Aircraft Corporation Pty (CAC) carried out licence manufacture of the Sabre. The aircraft was designated CA-26 by the manufacturer and the RAAF designated it Sabre Mk 30. The prototype Avon-powered Sabre (A94-101) first flew on 3 August 1953. Flt Lt Bill Scott of the RAAF was loaned to Commonwealth Aircraft to carry out the initial flight test programme, which, considering the changes, went remarkably well. The first production Sabre Mk 30 (CA-27) made its first flight on 13 July 1954 and it was delivered to the RAAF on 30 August. The first batch of 21 Sabre Mk 30s was powered by the Rolls-Royce Avon engine, but follow-on aircraft were powered by the CAC-built Avon Mk 20, which had a constant-taper Solar jet pipe. Twenty-one Avon 20-powered Sabres were designated Sabre Mk 31 and the Mk 30 aircraft were later brought up to Mk 31 standard. These were followed by 69 Sabre Mk 32s, which were fitted with the F-86F's '6-3' wing, uprated Avon Mk 26 engines and four underwing hardpoints for stores.

The first 12 Mk 30s were issued to the Sabre Conversion Flight of No. 2(F) OTU at Williamtown, NSW. The Avon Sabre equipped Nos 3, 75 and 77 Squadrons. On 11 November 1958, No. 3 Squadron was deployed to Butterworth, Malaya, as part of the Commonwealth Strategic Reserve, being joined there on 1 February 1959 by No. 77 Squadron. Together with the Canberras of No. 2 Squadron, they formed No. 78 RAAF Wing, which was used to provide offensive support for the Commonwealth Air Forces in the campaign against communist terrorists in Malaya. No air strikes were necessary until August 1959, when the Sabres strafed several CT camps in Northern Pahang. When the Malayan Emergency ended the RAAF Sabres returned to their primary role of air defence, a task they shared with the Gloster Javelins and Hawker Hunters of the RAF. However, in the early 1960s a period of confrontation between Malaysia and Indonesia

CL-13B Sabre 6 23707 served with No. 434 Squadron, No. 3 Fighter Wing, RCAF, based at Zweibrücken, Germany, from March 1953 to 1963. *(Canadair)*

began. When, in September 1964, armed Indonesian regular troops began making incursions into the Malay Peninsula, HQ Far East Air Forces put all its strike/attack units on alert. On 28 October 1964, the RAAF Sabres took part in a full-scale air defence exercise. With the end of confrontation in 1966, No. 78 RAAF Wing provided detachments of Sabres to Labuan Island off the coast of Sarawak, to relieve the Hunters of No. 20 Squadron which had provided air support for ground forces operating along Malaysia's disputed Borneo frontier with Indonesia. By now, RAAF Sabres were each armed with two Sidewinder AAMs mounted on the inboard underwing stores pylons. The Sidewinder installation had been successfully tested on Sabre A94-946 in 1959 and a supply of missiles had been flown direct from the USA in two C-130A transports of No. 36 Squadron, RAAF, in February 1960. On 1 June 1962, No. 79 Squadron became the fourth RAAF Sabre Mk 32 unit when it reformed at Singapore. Supported by C-l30As of No. 36 Squadron, the Sabres flew to Ubon, Thailand as part of a South East Asia Treaty Organization (SEATO) deployment and remained there until August 1968, when they redeployed to Butterworth. No. 79 Sqn disbanded in July 1968.

3es

Meanwhile, early in 1964, the Sabres of No. 75 Squadron at Williamtown were replaced by Dassault Mirage IIIO(A) aircraft. In May 1967 they flew to Butterworth to replace No. 3 Squadron, which had returned to Williamtown in February to convert to the Mirage. In February 1969, No. 3 Squadron returned to Butterworth where it replaced No. 77 Squadron, which returned to Williamtown to convert to the French fighter.

Bangladesh

The Bangladesh Biman Bahim/BDF (Bangladesh Defence Force) acquired five Canadair Sabre 6s from Pakistan in 1971 and operated them until late in 1973, when 12 Soviet-supplied MiG-21MF fighters became fully operational.

Bolivia

In October 1973 the *Fuerza Aérea Boliviana*/FAB (Bolivian air force) acquired nine F-86F-30 Sabres from Venezuela, to equip 1 Grupo Aéreo de Caza (later Grupo Aéreo de Caza 32) at Santa Cruz/El Trompillo. Two were lost in 1978 over the Amazon jungle and more were lost when a light aircraft hit the FAB hangar at Santa Cruz in July 1984. In August 1983, a lack of Argentinian maintenance support resulted in three more Sabres being withdrawn from service for a time. By 1987 four Sabres were on strength at Santa Cruz and they soldiered on until 1992, when one of the aircraft exploded in flight and the three surviving aircraft were withdrawn from the Bolivian inventory.

Canada

Early in 1949, the Canadian government selected the F-86A Sabre to form the nucleus of a new RCAF jet fighter force and Canadair Ltd of Montreal was selected to build the aircraft under licence at Cartierville. From the outset, the RCAF intended installing the Avro Canada Orenda engine in the Sabre as soon as it was available. In the meantime, in August 1949 a contract was signed for 100 J47-GE-12 powered F-86As, which specified that the first aircraft was to be delivered within 12 months. On 9 August 1950 the first (and only) F-86A (CL-13/Sabre Mk 1, 19101) which was

produced using NAA-made components, was flown for the first time by Alexander J. Lilly at Dorval Airport. (By this time NAA had switched production to the much-improved F-86E and Canadair followed suit – the Canadair-built F-86E-6 being designated CL-13/Sabre Mk 2.) Some 350 Sabre 2s were built. In February 1952 the USAF purchased 60 Sabre 2s from Canadair and after modification at the Fresno modification centre in California, they were delivered to the USAF (as 52-2833/2892) between April and July to equip the 4th and 51st FWs in Korea. Three others (19378, 19384 and 19404) went to the RAF (though 19738 was subsequently returned to Canada and replaced by a Sabre 4). The Mk 2 had entered service with the RCAF in spring 1951 and final deliveries were made in November 1952. In Europe, the RCAF Sabre 2 was the only swept-wing fighter in NATO. After their retirement, in 1954–55, Greece and Turkey each received 107 ex-RCAF Sabre 2s under MDAP.

Meanwhile, an early Orenda 3 turbojet was installed in an NAA-built F-86A-5 airframe (49-1069) which was then designated F-86A/O. In October 1950 this aircraft, flown by Maj. Robert L. Johnson, USAF, began high-altitude, high-speed flight testing at Edwards AFB, California. He then flew the Sabre to Canada. Canadair modified the 100th CL-13 Sabre 2 (RCAF 19200) to receive an Orenda 3, which became the only Sabre 3. Glen Lyons flew it for the first time on 4 June 1952. It was intended to install the Orenda in the Sabre 4 but this was postponed. In 1952 the J47-GE-12 was replaced by the 27.13-kN (6,100-lb) thrust J47-GE-27. The modified aircraft was designated CL-13 Sabre 4 and was flown for the first time on 28 August 1952, by Hedley I. Everard. In all, 438 Mk 4s were built, but only Nos 422 and 444 Squadrons operated them in RCAF service. Almost all were transferred to the RAF under MDAP and they equipped 11 squadrons, the vast majority in West Germany, remaining in RAF service until replaced by the Hawker Hunter by June 1956. At that time the remaining Sabre 4s were passed on to Italy and Yugoslavia.

In 1953 after about 800 Sabres had been built, the

28.17-kN (6,335-lb) thrust Orenda 10 turbojet was introduced to the Canadair production line (the F-86F's 'six-three' wing modification was also incorporated) and the Mk 5 Sabre was born. Just as with the RAAF's Avon-powered Sabre, the Canadian Sabre needed some redesign of the airframe to accommodate the Orenda, but unlike the Avon, modifications to accommodate the slightly larger diameter Orenda were relatively straightforward. The first Orenda-powered CL-13A Sabre Mk 5 was flown on 30 July 1953 by W. S. Longhurst. A total of 370 Sabre 5s was built and all were delivered to the RCAF. When the Sabre 5s were replaced by Mk 6s, 60 Sabre 5s were given to West Germany.

The CL-13B Sabre Mk 6 was the final Canadian Sabre version. It was powered by the Orenda 14, which weighed about 136 kg (300 lb) less than the Orenda 10, and featured a two-stage turbine producing 33.80 kN (7,600 lb) thrust. The result was a dramatic increase in high-altitude performance over the Mk 5, the Sabre 6 having an initial climb rate of 3597 m (11,800 ft) per minute and an operational ceiling of just over 15240 m (50,000 ft). It could also operate in the ground-attack role, carrying a 1814-kg (4,000-lb)

F-86F-30 52-4441, of The Republic of China Air Force (RoCAF), was one of a batch of 859 (NA-191) models. Formosa (now Taiwan) was the largest overseas user of Sabres, receiving 320 F-86Fs and seven RF-86Fs in the period 1954 to 1958, as well as a number of follow-on F-86Ds and Fs. *(via Robert Jackson)*

underwing load of bombs, rockets or fuel tanks. W. S. Longhurst flew the Sabre 6 for the first time on 19 October 1954 and the first production aircraft was completed on 2 November. In all, 647 Sabre 6 aircraft were built. The RCAF received 382 examples, South Africa 34, West Germany 225 and six were delivered to Colombia, in June 1956. Mk 6 production brought the total production run of Canadair-built Sabres to 1,815. RCAF Sabre 6s served with nine squadrons in West Germany. Sabre 5s and 6s were used by the RCAF 'Golden Hawks' aerobatic team.

With war in Korea and rising East-West tension in Europe, on 1 November 1951, No. 1 Fighter Wing, RCAF (NATO), was formed at RAF North Luffenham, Rutland, UK. Its mission, together with the F-86A Sabres of the 81st FW, USAF, based at RAF Shepherd's Grove and RAF Bentwaters, Suffolk, was the air defence of Great Britain. Nos 410, 419 and 441 Squadrons, No. 1 FW, remained at North Luffenham for the next three years, at which time UK air defence was undertaken by the Sabres and Hunters of RAF Fighter Command.

On 1 October 1952, No. 1 Air Division, RCAF, had become Canada's air contribution to NATO in western Europe. Ultimately it would consist of 12 squadrons of Sabres, in four fighter wings. In 1951, No. 1 FW shipped its Sabres of Nos 410 and 411 Squadrons to NATO, with No. 439 Squadron flying the Atlantic to RAF North Luffenham, England, in 1952, in Operation *Leapfrog 1*. In October that same year the 48 Sabres of Nos 416, 421 and 430 Squadrons, No. 2 FW, deployed from Canada to Gros Tenquin, France, via Greenland, Iceland and the UK, in Operation *Leapfrog 2*. In March 1953, Nos 413, 427 and 434 Squadrons, No. 3 FW, similarly flew to Zweibrücken in West Germany, while Nos 417, 422 and 444 Squadrons, No. 4 FW, deployed to Baden-Söllingen. The Canadian wings, together with USAF Sabre wings and the French 1st Commandement Aérien Tactique (CATAC), formed part of the 4th Allied Tactical Air Force. The 4th ATAF Sabre squadrons, like their 2nd ATAF counterparts in the north, took turns at maintaining the NATO Zulu alert commitment, which meant keeping a battle flight of two Sabres from each wing on immediate standby and the remainder at 15 minutes' readiness. In time of war, the entire fighter force had to be airborne within one hour of the alert.

In the summer of 1954, No. 1 Air Division took part in Exercise *Carte Blanche*, the largest NATO air defence exercise held up to that time. The Sabres were part of the 'Southland Air Force' and flew about 2,000 sorties during the exercise. On 14 November 1954, No. 1 FW at North Luffenham redeployed to the continent. No. 410 Squadron deployed to Baden-Söllingen and on 20 December, No. 441 Squadron moved to Zweibrücken, while on 31 March 1955, No. 419 Squadron relocated to Marville. Twice a year the RCAF Sabre pilots flew to Rabat, Morocco, for armament practice camps (APCs) using a *Leapfrog* type exercise and two Bristol Freighters as support aircraft. These APCs continued until 1957, when the squadrons began using a new Armament Practice Range at Decimomannu, Sardinia. When, in September 1957, Exercise *Counterpunch* took place, No. 1 Air Division was equipped with the Sabre 6. That same year Nos 410, 413, 416 and 417 Squadrons returned to Canada and Nos 400, 419, 423 and 440 Squadrons equipped with Avro Canada CF-100 all-weather fighters, replaced them.

In July 1959, the Canadair-built CF-104 Starfighter was selected to replace the Sabre in No. 1 Air Division. By 1961, the Sabre 6s were beset with fatigue problems and in March all air fighting manoeuvres were curtailed. In August 1962, No. 427 Squadron disbanded and reformed as No. 1 Air Division's first CF-104 unit. Early in 1963, re-equipment of all RCAF Sabre squadrons followed. The last RCAF Sabres in Europe were retired on 14 November 1963, when No. 439 Squadron at Marville, France, was stood down.

Colombia

Six Canadair Sabre 6s were purchased by Colombia in May 1956 and were used to equip 1° Grupo de Combate of the *Fuerza Aérea Colombiana* (FAC). In 1963 the FAC received up to three ex-*Ejército del Aire* (Spanish air force) F-86F Sabres. All surviving Sabres were retired in 1966.

Republic of China

The Republic of China Air Force (RoCAF) on Formosa (now Taiwan), was the largest overseas user of Sabres, receiving 320 F-86Fs and seven RF-86Fs in the period 1954 to 1958, as well as an indeterminate number of follow-on F-86Ds and Fs. The first F-86Fs were delivered in November 1954 and by June 1956 the RoCAF had received 171 F-86Fs. These equipped the 2nd, 3rd and 5th FWs between 1955 and 1956 and RF-86Fs were issued to the 12th Squadron, 5th FW in 1955. Sabres were involved in several combats and interceptions with Red Chinese MiGs during the 1950s and 1960s, including the Formosa Crisis of 1958, when Sabres armed with Sidewinder AAMs fought frequent air battles with Communist Chinese MiG-17s. The RoCAF claimed 31 MiGs shot down for the loss of only two F-86s, but in reality only six Sidewinders were fired during the crisis and these brought down four MiG-17s. The F-86Ds were retired in 1966 and from 1960 onwards the F-86Fs were gradually replaced by Northrop F-5s and F-104 Starfighters. The last RoCAF Sabre unit was the 2nd Fighter Wing at Hsinchu, which replaced its F-86Fs with F-100As in 1971.

Denmark

Beginning in June 1958 the *Kongelige Danske Flyvevaben* (Royal Danish Air Force, RDAF) received 38 ex-USAF F-86D-30 and -35 aircraft to replace the Meteor NF.Mk 11s of Eskadrille (Esk) 723 at Aalborg. Esk 726 became the second F-86 unit later that same month and moved to Aalborg to form an all-Sabre base. In May 1960 the RDAF received a second batch of 21 F-86Ds to convert Esk 728, which replaced its F-84Gs at Skrydstrup. Service use continued until June 1965, when Esk 723 and 726 converted to the F-104G Starfighter. Esk 728 disbanded on 31 March 1966.

Ethiopia

During July and September 1960 a number of aircraft, including 12 F-86F Sabres, were delivered to the Imperial Ethiopian Air Force (IEAF) as part of a Mutual Assistance Agreement with the United States. The Sabres were used by the Ethiopian

RAF CL-13 Sabre 4s XB927, XB928, XB916 and XB953 are shown in formation. *(Flight)*

government on 15-16 December 1960 against rebel positions in the capital, Addis Ababa, during an attempted coup by the Swedish- (UN-) trained Imperial Guard. In late 1961 Ethiopia sent four of the Sabres to the Congo, to assist the UN peacekeeping forces there. In the summer of 1962, two more IEAF F-86Fs were dispatched to the troublespot. After a dispute, the Ethiopian government pulled its F-86Fs out of the Congo in October 1962 and they returned to Ethiopia under the pretext of a long-range navigation exercise. Two years later, the IEAF Sabres were used in the Horn of Africa war with Somalia (and it is rumoured they may have also taken part against Somalia in the 1977-78 war over the disputed Ogaden Desert region). During 1970 it is believed that just over 20 F-86Fs were acquired by Ethiopia from the Imperial Iranian Air Force. In 1977 the remaining dozen or so F-86Fs were withdrawn from service.

France

In 1955 the French government announced that as a stopgap measure after delays into service of the indigenous Mirage IIIC and Vautour IIN aircraft, the *Armée de l'Air* would receive F-86K Sabres. 13ᵉ

Escadre de Chasse Tout Temps (ECTT, All Weather Fighter Wing) was formed on 1 March 1955 at Lahr AB in West Germany to receive them, but, ironically, the delivery of the first of 60 Fiat-built F-86Ks did not take place until 4 September 1956. On 1 October 1956 Escadron 1 'Artois' and Escadron 2 'Alpes' were formed within 13ᵉ ECTT and the Sabres were finally assigned. On 1 April 1957, 13ᵉ ECTT moved from Lahr to a new purpose-built base at Colmar-Mayenheim, France. In 1959 the remaining Sabres were returned to Fiat for overhaul and modification and received extended leading edges and increased span. In June 1960, Fiat fitted Sidewinder launch rails to at least 29 of the remaining F-86Ks. In January 1962 Escadron 1/13 'Artois' began Mirage conversion and its F-86Ks passed to Escadron 2/13 'Alpes', which relinquished its Sabres that April, although as an interim measure EC 3/13 was formed to operate the F-86Ks until the Mirage conversion programme was complete. Early in 1962, 22 of the F-86Ks were transferred to the AMI, the last F-86Ks in Armée de l'Air service being phased out in October that same year.

Great Britain

In the spring of 1952 it was announced that pending the arrival of the Hawker Hunter, the RAF would receive Canadair-built CL-13 Sabres (three Sabre 2s and 367 Mk 4s), which were ordered under MDAP. These aircraft, very similar to the F-86E-10 and powered by the 23.12-kN (5,200-lb) thrust J47-GE-13 engine, were urgently needed to replace the aging de Havilland Vampire FB.Mk 5 and FB.Mk 9 in ten squadrons of the 2nd Tactical Air Force in West Germany. In November 1952, a further 60 Sabre 4s were offered and accepted to replace the Gloster Meteor F.Mk 8s of Nos 66 and 92 Squadrons, RAF Fighter Command, at Linton-on-Ouse, Yorkshire.

Instead of making deliveries to Britain by sea, to obviate the prohibitive cost of dismantling and reassembling the aircraft, all the 430 Sabres were flown the 4828 km (3,000 miles) across the Atlantic in convoys, in Operation *Becher's Brook*. This operation drew upon the experience gained during Operation *Leapfrog*, in which RAF pilots flew as observers. The first three Sabres were designated Mk 2s and were used to form the Ferry Conversion Unit under the control of No. 1 RCAF Wing at North Luffenham. The rest were all Mk 4s,

An F-86F-30 Sabre of No. 2 Squadron, SAAF, and 67th FBS F-86Fs of the 18th Fighter Bomber Wing, taxi out at K-55 Osan in the late spring of 1953. *(Dick Kempthorne via Larry Davis)*

These F-86F-30 Sabres hailed from No. 2 Squadron, SAAF, 18th Fighter Bomber Wing and are seen taking off from K-55 Osan in 1953. The 'Flying Cheetahs'' first Sabre sorties were flown on 16 March 1953. No. 2 Squadron's Sabres had flown over 2,000 sorties by the end of the Korean War. *(via Larry Davis)*

North American F-86A-5-NA Sabre, 335th FIS, 4th
FIW, Korea

North American F-86A-5-NA Sabre, 78th FIS, Korea

North American F-86A-5-NA Sabre, 116th FIS,
Shepherd's Grove, England, 1951

North American F-86E-10-NA Sabre, 336th FIS, 4th
FIW, Korea, Capt. Robert Risner, eight kills

North American F-86E-10-NA Sabre, 334th FIS, 4th
FIW, Korea, Maj. C. Blesse, ten kills

North American F-86E-10-NA Sabre, 16th FIS, 51st
FIW, Korea

North American F-86E-1-NA Sabre, 25th FIS, 51st
FIW, Korea

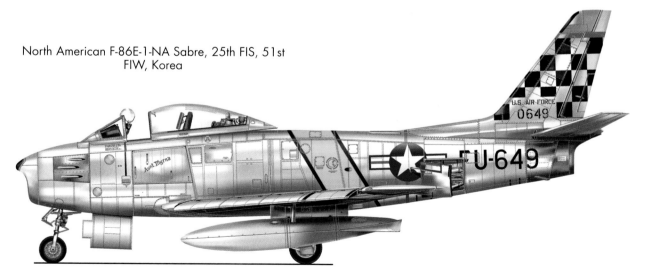

Above: CL-13B Sabre 6 23757, of the RCAF, is shown in flight. *(Canadair)*

Below: Sabre 5s and 6s were used by the RCAF's 'Golden Hawks' aerobatic team. *(RCAF)*

which had a modified cabin air conditioning system and other minor modifications. The RAF ferry pilots received conversion training on the F-86E with No. 1 FW, RCAF at North Luffenham, Leicestershire, before joining No. 1 Long Range Ferry Unit (later No. 147 Squadron) in Canada, where they underwent further training in extreme cold, and bad weather and Arctic survival. *Becher's Brook* started on 8 December 1952, when the first convoy of 12 Sabres (which included the three

F-86F-30 52-4458, of the Chinese Nationalist Air Force, was at Taipei in March 1965. *(Harry Holmes)*

Mk 2 aircraft) left for the UK from RCAF Bagotville, Quebec. The average airborne time for the Atlantic crossing was 6 hours, but crossing times could be anything from four days to three weeks, depending on the weather. The cruising altitude flown was between 9144 and 10668 m (30,000 and 35,000 ft) and the intense cold at these altitudes produced aileron control locking, problems caused by the contraction of the aileron torque tubes. Landings were therefore made without using aileron. During the early movements, No. 147 Squadron began its ferry flights from Bagotville and flew via Goose Bay in Labrador, across the North Atlantic via Bluie West 1 (Narsarssuak) in southern Greenland, Keflavik, Iceland, and Prestwick, Scotland, before dispersal to maintenance units in the UK. Later movements began from St Hubert near Montreal and ended at

Kinloss, Scotland. For a short period RNAS Lossiemouth was used as the arrival airfield.

Altogether, No. 147 Squadron flew ten transatlantic Sabre convoys, each ranging in number from 11 to 54 aircraft, while on two occasions, to clear a build-up of waiting Sabres, a double-shuttle system was used and 64 aircraft were ferried by 32 pilots who returned in Hastings aircraft for the second leg of the ferry flight. The last convoy arrived in Scotland on 19 December 1953. Five Sabres were lost during Operation *Becher's Brook*.

Sabres equipped No. 229 Operational Conversion Unit at RAF Chivenor, Devon, to provide fast-jet experience for squadron pilots, while a few were issued to the Central Gunnery School, the Fighter Weapons School and the Air Fighting Development Squadron. The first to equip with the Sabre 4 were the three former Vampire FB.Mk 5 squadrons of the Wildenrath Wing of the 2nd TAF in West Germany, when in May 1953, Nos 3 and 67 Squadrons became the first swept-wing fighter squadrons in the RAF, No. 71 Squadron receiving its Sabres in October. A Sabre Conversion Flight was also established at Wildenrath to supplement the England-based No. 229 OCU. During the period 1953 to 1954 seven

CL-13 Sabre 4 19523 (ex-RAF XB620), was seen with the 4° Aerobrigata Intercettori Diurni, AMI at Capone on 2 June 1967. *(Harry Holmes)*

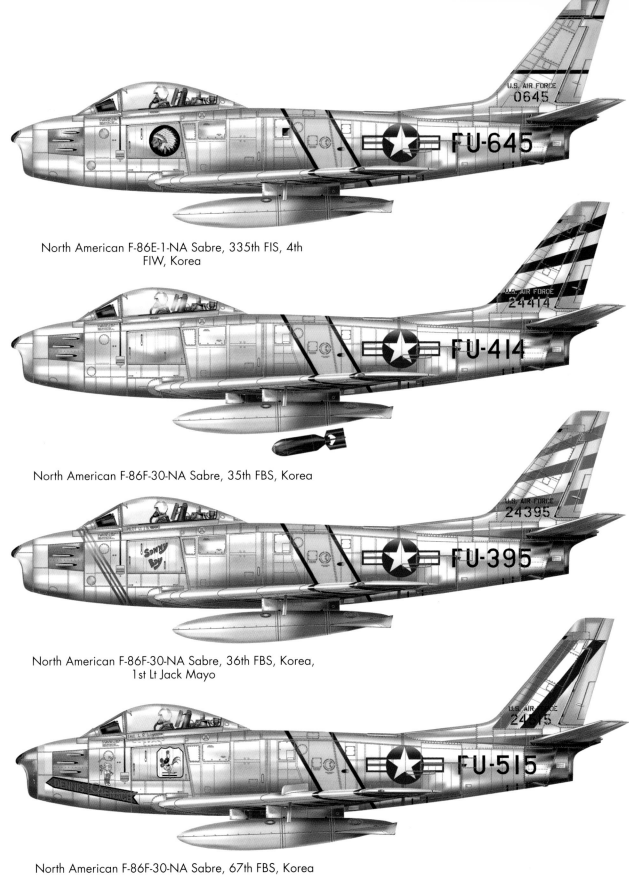

North American F-86E-1-NA Sabre, 335th FIS, 4th
FIW, Korea

North American F-86F-30-NA Sabre, 35th FBS, Korea

North American F-86F-30-NA Sabre, 36th FBS, Korea,
1st Lt Jack Mayo

North American F-86F-30-NA Sabre, 67th FBS, Korea

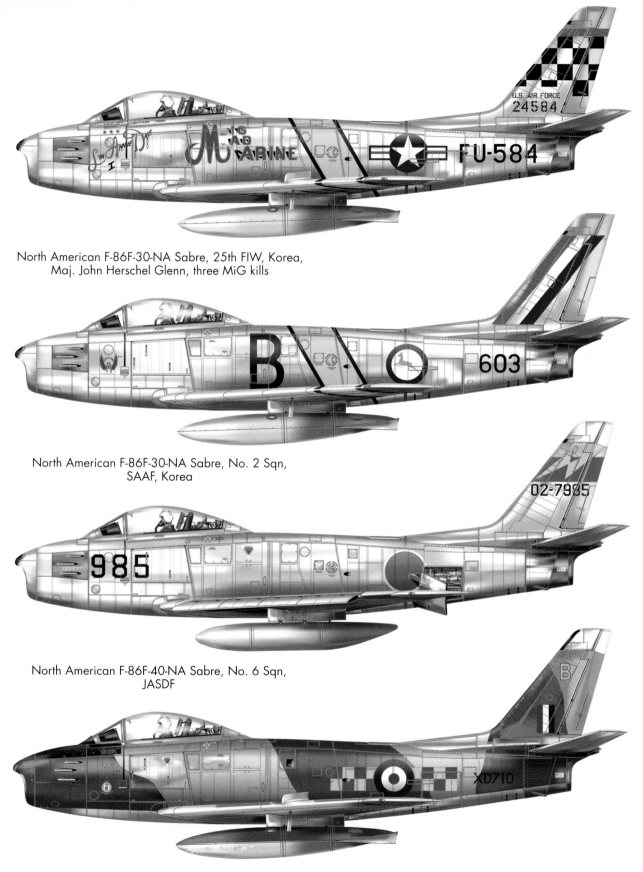

North American F-86F-30-NA Sabre, 25th FIW, Korea,
Maj. John Herschel Glenn, three MiG kills

North American F-86F-30-NA Sabre, No. 2 Sqn,
SAAF, Korea

North American F-86F-40-NA Sabre, No. 6 Sqn,
JASDF

Canadair Sabre Mk 4, No. 92 Sqn, RAF

more RAF squadrons – No. 4 at Jever, Nos 20, 26 and 234 at Oldenburg, No. 93 at Alhorn and Nos 112 and 130 at Brüggen had also re-equipped with the Sabre 4. In England meanwhile, No. 66 Squadron was the first to exchange its Meteor F.Mk 8s for Sabre 4s, in December 1953, and No. 92 Squadron similarly re-equipped in February 1954. These two squadrons received 60 Sabres, enough

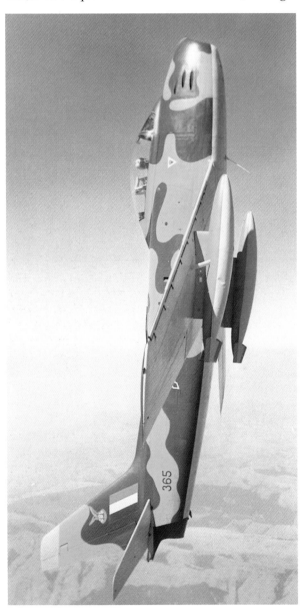

for 100 per cent reserves. (Some 18 Sabres were lost in accidents during their service.)

During 1953–54, de Havilland Venom FB.Mk 1s began equipping two RAF Wings in 2nd TAF. In 1954 2nd TAF lost 23 Sabres, roughly half the accidents being due to aircraft or systems failure. In October, after a Sabre broke up in mid-air, all 2nd TAF Sabres were grounded for several weeks for structural checks and flying operations did not return to normal until early in 1955. In April 1954 the Hawker Hunter F.Mk 4 began replacing the Venom in Nos 98 and 118 Squadrons, which moved to Jever, where they operated alongside the Mk 4 Sabres of Nos 4 and 93 Squadrons in a four-squadron air defence wing. In May 1955, No. 14 Squadron at Fassberg exchanged its Venoms for Hunters and moved to Oldenburg to form a three-squadron air defence wing with the Mk 4 Sabres of Nos 20 and 26 Squadrons. In June, No. 26 Squadron became the first Sabre squadron to convert to the Hunter. It was followed in July by No. 4 Squadron at Jever and in November by No. 20 Squadron. In January 1956, No. 67 Squadron at Brüggen and No. 93 Squadron at Jever converted to the Hunter F.Mk 4. In April the Brüggen Wing became an all-Hunter wing with the conversion of Nos 112 and 130 Squadrons, while No. 71 Squadron at Wildenrath also received Hunters that same month. In May 1956, Nos 3 and 234 Squadrons at Geilenkirchen became the last two squadrons in 2nd ATAF to convert to the Hunter. No. 147 Squadron's pilots collected the now redundant Sabres from their German bases and flew them to RAF Benson, Oxfordshire for refurbishment prior to service with other NATO air forces and other air arms. By 21 June 1956 some 267 Sabres had been flown back to Britain, prior to transfer to USAF control. In England meanwhile,

CL-13B Sabre 6 365 is shown in the colours of the SAAF, which in 1956 received 34 Sabre 6s to equip No. 1 Squadron at Pietersberg and No. 2 'Cheetah' Squadron at Waterkloof. In June 1963, No. 2 Squadron began re-equipment with the Mirage IIICZ and its Sabres were re-assigned to No. 1 Squadron, which did not give up its Sabres until 1977, when it re-equipped with the Mirage F1AZ. *(SAAF)*

Nos 66 and 92 Squadrons used the Sabre until spring 1956 when they re-equipped with Hunter F.Mk 4s. Between 1956 and 1958, 302 Sabres were returned to USAF charge. XB982 was acquired by Bristol Siddeley as a test-bed for the Orpheus 12 engine in the period 1958 to 1959.

Greece

In July 1954, the *Elliniki Polemiki Aeroporia* (Royal Hellenic Air Force) received the first of 107 refurbished Canadair Sabre 2 aircraft to form a day interceptor component. These went to re-equip the No. 112 Pterighe (Wing) at Elefsina AB, No. 341 Day Fighter Squadron (Mira) being the first unit to so equip. In 1955, follow-on deliveries enabled Greece to equip Nos 342 and 343 Mire (Squadrons). In August 1958, No. 341 Mira formed an aerobatic team, the 'Hellenic Flame', with five (later increased to seven) F-86E Sabre 2s and displayed in West Germany, Italy, Turkey and France, before disbanding in September 1964. No. 341 Mira converted to the F-5 in 1965 and No. 342 Mira did likewise in January 1966. In May 1960, the first of 35 F-86Ds were delivered and they began equipping No. 337 Mira at Elefsina. Next to convert to the F-86D was No. 343 Mira, which began conversion from the F-86E(M) in May 1961 and operated this type until September 1965. No. 337 Mira finally retired its Sabre Dogs in May 1967.

RF-86F-30 52-4492 of the 32nd Squadron, Republic of Korea Air Force (RoKAF) was seen at Suwon on 1 May 1968. Ten ex-USAF RF-86Fs were delivered to the RoKAF from 11 April to 2 May 1958. *(Harry Holmes)*

Honduras

In 1976 Honduras purchased eight ex-Yugoslav air force Canadair CL-13 Mk 4 (F-86E(M)) Sabres as a stopgap pending delivery of Dassault Super Mystère B.2s from Israel. As it turned out, the CL-13 Mk 4s and the Super Mystères arrived almost at the same time and so the Sabres were considered superfluous by the *Fuerza Aérea Hondureña* (FAH). Four were soon presented to the Venezuelan government, the remaining Sabres soldiering on in FAH service until their retirement in 1986.

Indonesia

In December 1972 Australia presented the *Angkatan Udara Republik Indonesia* (AURI, Indonesian air force) with the first of 12 ex-RAAF CAC-built Sabre Mk 32s, which were used to re-equip No. 14 Day Fighter Squadron at Iswahyudi AB in East Java. In 1974, the AURI was renamed Tentara Nasional Indonesia-Angkatan Udara (TNI-AU) and No. 14 Squadron was renamed Sat-Sergap F-86 Squadron. In 1976, Indonesia received five more Sabres which were transferred from the Royal Malaysian Air Force. In July 1978 an

Canadair Sabre Mk 5, *Luftwaffe*

North American F-86H-10-NA 'Hog' Sabre, Arizona
Aerospace Foundation

North American F-86H-10-NA 'Hog' Sabre, 101st TFS,
Massachusetts Air National Guard

North American F-86D 'Dog' Sabre, 97th FIS, Wright-
Patterson AFB, 1950s

North American F-86D 'Dog' Sabre, 25th FIS,
Okinawa, 1950s

North American F-86L 'Dog' Sabre, 331st FIS

North American F-86K Sabre, 337 Skv, KNL

North American F-86F Sabre, warbird in the colours of
the 25th FIS

Above: This ex-JG74, Fiat-built F-86K 0960, was marked for the *Fuerza Aérea Venezolanas* (Venezuelan Air Force) on a test flight by Dornier of Munich on 1 September 1967. *(Harry Holmes)*

Below: F-86K JD-302, of the Luftwaffe's JG74, flies near Neuburg on 18 May 1965. *(Harry Holmes)*

Indonesian Sabre aerobatic team was formed. The last Indonesian Sabres were retired late in 1980.

Iran

In 1960 the *Nirou Havai Shahanshahiye Iran* (Imperial Iranian Air Force, IIAF) received 52 ex-USAF F-86F-25 and F-86F-30 aircraft courtesy of MDAP, which were refurbished by Fiat in Italy en route. In 1960 the IIAF 'Golden Crown' aerobatic team converted to the Sabre from the F-84G and operated six Sabres until they were replaced by the F-5 in 1966. In 1963, the IIAF allocated four F-86Fs to UN operations in the Congo. In February 1966 Iran revealed that 80 ex-Luftwaffe Canadair Sabre Mk 6s had been purchased, ostensibly to form a day-fighter wing in Iran, but they were really intended for the Pakistan Air Force. The Iranian Sabres were finally retired in about 1971.

Iraq

The Iraqi air force (IAF) obtained five F-86F Sabres in July 1958, but none entered operational service. Following the coup which toppled King Faisal II, America made no further deliveries of F-86Fs or any other aircraft.

Left: F-86E-10 51-2845, of the 62nd FS, 56th FG, was based at O'Hare International Airport, Chicago, in June 1953. *(Picciani via Larry Davis)*

Below: F-86Fs 51-12963 and 51-12954, of the 4th FIW, were en route to the Yalu River in October 1952. *(via Larry Davis)*

F-86D-60(L) 53-3681, of the 191st FIS, Utah ANG, at Salt Lake City on 18 May 1965. *(Harry Holmes)*

Above: F-86A-5 49-1157, of the 3596th CCTS 'Cadillac Flight', was photographed at Nellis AFB in 1952. *(via Larry Davis)*

Left: F-86F 52-4608 is seen at Edwards AFB. This Sabre (as N57963) is now owned by Robert D. Scott, at San Martin, CA. *(via Larry Davis)*

Below: F-86Es of the 51st FIW line up at K-13 in 1953. *(via Larry Davis)*

Armourers work on an F-86F-2 *Gun-Val* at K-14 in 1953. *Gun-Val* was introduced after requests from pilots for heavier-calibre guns, while still retaining most of the rate of fire of the .50-cal guns. Four F-86E-10s and six F-86F-1s were each armed with four T-160 20-mm guns and re-designated as F-86F-2s, while two F-86F-1s armed with Oerlikon-type 20-mm guns were redesignated as F-86F-3s. Eight F-86F-2s were issued to the 4th FIW in March 1953 for 16 weeks of combat tests. *(Paul Peterson via Larry Davis)*

F-86H-10 (NA-203) 53-1296 *Lind 9th*, was the personal aircraft of Lt-Col David F. McCallister (ex 20th FG, 8th AF) of the 142nd FIS, Delaware ANG, New Castle County Airport, Delaware, on 6 June 1957. *(Harry Holmes)*

Right: Groundcrew change a 4th FIW Sabre engine at K-14 Kimpo late in 1951. *(via Larry Davis)*

Below: Groundcrew prepare F-86E-1 50-602 *Miss B*, of the 16th FIS, 51st FIW, for a combat mission to 'MiG Alley' in April 1952. *(via Larry Davis)*

Below: 18th FBG armourers, at K-55 Osan in 1953, unload 1,000-lb (454-kg) M44 GP bomb casings beside F-86F-30 52-4337 *The Stinger*, which was assigned to the 12th FBS, 18th FBW. Note the SAAF orange, white and blue stripes on the rudder. *(Bill Grover via Larry Davis)*

CL-13 Sabre 2 (F-86E-6) 19448, was one of 350 built for the RCAF. After its retirement, it became one of 107 examples acquired under MDAP for the Elliniki Polemiki Aeroporia, beginning July 1954 to form a day-interceptor component. *(RHAF)*

Italy

During 1956–57, the AMI acquired 180 ex-RAF Canadair Sabre Mk 4s (F-86E(M)s) to equip the 2° and 4° Aerobrigata Intercettori Diurni (fighter day wings). In 1956 the 4° Aerobrigata formed the AMI's first Sabre aerobatic team, the 'Cavalline Rampante' ('Rampant Horses') at Pratica del Mare. Four aircraft were operated for two seasons in the period 1957 to 1959, until they were replaced by the F-84Fs of the 6° Aerobrigata. During 1958–59, six Sabre 4s of the 2° Aerobrigata at Cameri formed the 'Lanceri Neri' ('Black Lancers'). Early in 1961 they were superseded by the famous 'Frecce Tricolori', which operated with nine Sabre 4s until early 1964, when the team converted to the Fiat G.91. The last Fiat-built F-86K was delivered at the end of October 1957, although 22 F-86Ks were acquired from France in 1962. In the early 1960s,

the Canadair Sabre began to be phased out in favour of the F-104 Starfighter and the G.91R. The 2° Aerobrigata's 13° and 14° Gruppi (squadrons) re-equipped with the G.91R, while the 9° and 10° Gruppi in the 4° Aerobrigata re-equipped with the F-104G. In December 1971, 12° Gruppo converted from the F-86K to the F-104G and all remaining F-86Ks were transferred to 23°, now the last Sabre unit in the AMI, and which converted to the F-104S in March 1973.

Japan

When the *Nihon Koku Jieitai* (Japan Air Self-Defence Force, JASDF) was formed on 1 July 1954, its planned strength was seven fighter-interceptor and two all-weather interceptor wings (each with three Hiko-tais (squadrons)). To equip these units 526 F-86Fs and F-86Ds, as well as 54 RF-86Fs to

F-86D 84-8142 served with the JASDF. *(APN via Jerry Scutts)*

equip three tactical reconnaissance squadrons, were needed. A licence and joint production agreement was reached between Mitsubishi and NAA for the F-86F-40 model. In April 1956 an initial batch of F-86F-40s was shipped from the USA to Japan and by mid-1957 180 had been delivered. The JASDF, however, had insufficient qualified Sabre pilots, so 45 of the F-86F-40s were placed in storage and eventually returned to the USA in February 1959. Meanwhile, beginning in December 1955, 30 ex-FEAF F-86F-25 and -30 Sabres were delivered to the JASDF under MDAP for training purposes. The first of 300 F-86F-40s assembled by Mitsubishi flew on 9 August 1956 and all had been delivered by 25 February 1961. In October 1956 the 2nd Day Fighter Wing, the first tactical F-86F unit to form, was activated at Chitose. This unit comprised the 103rd, 201st and 203rd Hiko-tais. By February 1957, the 3rd (101st, 102nd and 105th Hiko-tais) at Matsushima and the 4th FW (Day) consisting of the 5th and 7th Hiko-tais, also at Matsushima, had been activated.

Also in 1957, the JASDF received four F-86D Sabres to train pilots for the nucleus of an all-weather fighter force. In January 1958, the first of

122 F-86Ds began arriving in Japan and by the end of 1962 had been used to equip four all-weather interceptor squadrons. Also in 1962, the first of 18 RF-86F Sabres were issued to the 501st Hiko-tai of Reconnaissance Command at Iruma. Among the other units to operate the Sabre was the 'Blue Impulse' aerobatic team, which formed in March 1960 and flew its last F-86F display on 8 February 1981, before conversion to the Mitsubishi T-2 began. In November 1961, the tenth and last F-86F Sabre Hiko-tai was activated. On 20 November 1965 2nd Hiko-tai disbanded and in December the 9th Hiko-tai followed suit. In December 1965, F-104 Starfighters at Hyakuri AB assumed the airborne defensive duties for Tokyo from the F-86F Sabres at Iruma. Between 1 October 1964 and spring 1966, seven F-86F Squadrons – the 201st, 202nd 203rd, 204th, 205th, 206th and 207th Hiko-tais – became operational with the F-104J Starfighter. Between 1 December 1967 and October 1968 the F-86D Hiko-tais also began to be withdrawn. In April 1977 the RF-86Fs were finally withdrawn. In 1979 the 6th Hiko-Tai, the last remaining F-86F squadron, converted to the Mitsubishi F-1. The Koku Sotai Shireibu Hiko-tai (HQ Squadron) was the last unit to give up its Sabres, flying the last F-86F sortie on 15 March 1982.

Republic of Korea

Following the end of the Korean War in 1953 and the subsequent withdrawal of most of the USAF units from the country, the Republic of Korea received US assistance in building up its Republic of Korea Air Force (RoKAF). To this end, a few F-86Fs were supplied in June 1955 and these were followed by more aircraft until delivery of the RoKAF's own aircraft. By 1956 the Republic of Korea had received 85 Sabres, the Sidewinder-armed F-86Fs being used to equip Nos 10 and 11 Wings in the fighter ground-attack role, and the all-weather F-86D interceptors equipping No 12 Wing. In 1958, the RoKAF received a further 27 F-86Fs and ten ex-USAF RF-86Fs. In 1960 these were augmented by the arrival of 45 F-86D aircraft, which were used to equip two interceptor squadrons in the 10th Wing. In 1978–79 the majority of F-86Fs and F-86Ds remaining were finally retired, although a few aircraft reportedly continued in service until the late 1980s.

Malaysia

In August 1969 the *Tentara Udara Diraja Malaysia* (TUDM, Royal Malaysian Air Force) received ten refurbished ex-RAAF Sabre Mk 32s; which were used to form No. 11 'Cobra' at Butterworth. In November 1971, a static training aircraft and six more Mk 32 Sabres were received. On 31 May 1975, No. 12 Kilau (Lightning) Squadron was formed and at the same time it absorbed No. 11 Squadron's surviving Sabres. In April 1976, No. 12 Squadron re-equipped with the Northrop F-5 and five TUDM Mk 32 Sabres were given to Indonesia.

Right: F-86F-35 53-1120, of Skvadron 336, Kongelige Norske Luftorsvaret is shown at Rygge. Skv 336 converted from the F-84G to the F-86F at Lista in 1957, and finally replaced its F-86Fs with Northrop F-5As in 1966. *(KNL)*

F-86K Q-278 (54-1278) of the KLu, was the second of the first batch of 15 aircraft shipped to The Netherlands on 1 October 1955 aboard USS *Tripoli*. Following assembly and test flight, No. 702 Squadron accepted 54-1277 and 54-1278 on 8 December 1955. Nos 700, 701 and 702 Squadrons formed on the F-86K between 1 August 1955 and 1 June 1956. *(KLu)*

The Netherlands

On 27 March 1953 the *Koninklijke Luchtmacht* (KLu, Royal Netherlands Air Force) was formed, to comprise six day fighter squadrons and three all-weather squadrons. In 1954 the Dutch government decided to equip the three all-weather squadrons with F-86K Sabres. Nos 700, 701 and 702 Squadrons formed on the F-86K in the period 1 August 1955 to 1 June 1956. On 1 January 1957, No. 702 Squadron was officially established as the OCU. All told, the KLu received 59 NAA-built F-86Ks by April 1957 and six Fiat-built examples arrived in April and May that same year. Some 13 F-86Ks were lost in flying accidents. From 1961, Sidewinder launch rails were retrofitted to most of the F-86Ks. In the summer of 1962 with the imminent introduction into service of the F-104G, a number of the KLu F-86Ks – known as the Kaasjager (K-fighter) – were withdrawn and after refurbishment passed to the AMI. No. 702 Squadron disbanded on 1 April 1962 and No. 701 was inactivated in 1963. On 30 June 1964, No. 700 Squadron disbanded and on 31 October 1964 the F-86K was officially withdrawn from service.

Norway

In September 1955 the *Kongelige Norske Luftorsvaret* (KNL, Royal Norwegian Air Force (RNorAF)) received 60 NAA-built F-86K Sabres. The first examples went to Gardermoen where they equipped Skvadron 337 in September 1955 and Skvadron 339 in July 1956. Between March 1957 and May 1958, the RNorAF received 90 F-86F Sabres under MDAP. Skvadron 332 at Rygge converted from the F-84G to the F-86F in April 1957. (Additionally, Skv 332 provided the Royal Norwegian Air Force's jet aerobatic team, the 'Flying Jokers Skvadrons'). Skvadrons 331 and 334 at Bodø and Skvadron 336 at Lista followed Skv 332 in converting to the F-86F. By mid-1960, the last F-84Gs had been phased out with the conversion of Skvadron 338 at Ørland. In May–June 1960, the KNL received six F-86F attrition replacements and a further 19 were received in January 1961. Meanwhile, in August 1960, Skv 334 converted to the F-86K. Then, in September 1963, Skvadrons 337 and 339 were disbanded, at which time they merged into Skvadron 332 and Skvadron 334 respectively. In April 1963 Skv 331 had replaced its F-86Fs with F-104G Starfighters. Skv 332 disbanded late in 1964 and in 1966 Skv 336 at Rygge replaced its F-86Fs with Northrop F-5As. Skv 334 retired the last F-86K Sabres in July 1967.

Pakistan

From 1956 to 1958, the Pakistan Air Force (PAF) received a total of 120 F-86F-40 and ex-USAFE F-86F Sabres from the US. The first deliveries of F-86Es in August 1956 went to Masroor where they equipped Nos 11 and 15 Squadrons and No. 2 Fighter Conversion Unit/Jet Conversion School. In June 1957 the first of many new-build F-86F-40s were used to equip the Jet Conversion School, now renamed No. 2 Fighter Conversion Squadron. From early 1958, F-86Fs were used to equip Nos 18 and 19 Squadrons and other PAF units. The PAF used Sabres in two wars with India, in 1965 and again in 1971. The first recorded Sabre victim was an Indian Air Force (IAF) Canberra, which was shot down on 10 April 1958 by a No. 15 Squadron F-86F Sabre pilot. On 1 September 1965, PAF Sabres claimed the destruction of four IAF Vampires. The so-called 'Seventeen Day War' began five days later when the PAF claimed seven IAF Hunters destroyed, and seven MiG-21s and five Mystères destroyed on the ground at Pathankot. On 7 September at Sargodha, the principal air base of the PAF, Sqn Ldr Mohammed Mahmud Alam, CO, No. 11 Squadron, claimed five Hunters shot down including four in 30 seconds. The events have since been well documented and it appears that two Hunters and not five were shot down. One of the victims suffered engine trouble and the Hunter flamed out due to fuel starvation. The pilot ejected and became a PoW. When a ceasefire ended hostilities on 23 September, the PAF had claimed at least 26 IAF aircraft downed in combat by Sabres alone, while admitting the loss of at least nine F-86Fs in combat and to ground fire. In fact, the PAF had lost 13 F-86Fs, seven of them in air combat.

In 1966 Pakistan used international subterfuge to acquire up to 80 ex-Luftwaffe Sabre 6s to rebuild the PAF after the 1965 war with India. To circumvent an arms embargo, which was still in force, Pakistan arranged for the Sabres to be delivered by asking its CENTO partner, Iran, to intervene. In July 1966, Iran admitted that the IIAF had received 80 Sabre 6s from Germany and had passed 60 of them on to Pakistan. In November 1971 India and Pakistan went to war again and the IAF's Hunter, MiG-21 and Sukhoi Su-7 aircraft hopelessly outclassed the PAF. Surprisingly perhaps, it was the highly manoeuvrable Hindustan Ajeet (licence-built Folland Gnat) that caused the PAF Sabre pilots the most concern in combat. The war ended in stalemate on 16 December and the PAF admitted to the loss of 34 aircraft of all types, but the total Sabre losses were probably nearer 49, some 15 of them on the ground. Nos 18 and 26 Squadrons were still using Sabre 6s in the late 1970s, but these aircraft were well past their fatigue life and after two broke up in mid-air, the rest were withdrawn.

Peru

In 1956 the *Fuerza Aérea del Peru* (Peruvian air force, FAP) received 12 F-86F Sabres under MDAP and they were used to equip No. 11 Escadrone de Caza (No. 11 Fighter Squadron) of Grupo de Caza 12 at Talara Limatambo. In 1960 a few Sabres were delivered to Peru as attrition replacements. (Between July 1963 and May 1965, the FAP lost four more Sabres.) The few remaining Sabres remained operational until 1976, when the FAP was re-equipped with Sukhoi Su-22 'Fitter-Cs'.

Philippines

In 1957 the *Hukbong Himpapawid ng Pilipinas* (Philippine air force, or PAF) received from the US some 30 F-86F-30s, a further 15 or more F-86F-30s following in June 1958 and another 18 or more in June 1959. A small number of ex-Spanish air force F-86s was received in April 1962. On 12 August 1960 the PAF received 20 F-86D-35s and later a few more 'Ds' were acquired. The F-86Ds of the 5th FW were phased out of service in July 1968. In

1972, 20 F-86Fs were transferred from Taiwan to the Philippines as part of MDAP, during a period of tension in the islands. In August 1965 the 6th TFS gave up its F-86s and converted to the Northrop F-5A, but those of the 7th and 9th TFSs continued to serve until the mid-1970s. Some may have soldiered on in a reserve unit until 1978.

Portugal

Beginning in August 1958 the *Força Aérea Portuguesa* (FAP, Portuguese air force) received the first of 65 ex-USAF F-86F-35 Sabres under MDAP. Esquadra 50 (later renumbered Esq 51) was formed on 4 February 1958 to work up on the Sabre at Basa Aérea BA2, at Ota, in the summer of 1959. Esquadra 52 became the second operational Sabre unit in October that same year and both squadrons combined to form Grupo de Caça 501. By 1979 the 15 surviving Sabres had been amalgamated into a single unit, Esquadra 201, which was based at Monte Real. Esquadra 52 disbanded on 12 June 1961 and its Sabres were absorbed by Esquadra 51. In 1962, the Portuguese Sabres were armed with AIM-9B Sidewinder AAMs for use in their role as Quick Reaction Alert (QRA) interceptors. From August 1961 a detachment of Sabres was based in Portuguese Guinea to forestall any insurrection there, and they took part in the fighting against rebels in Angola from the summer of 1963 to October 1964. The last Portuguese Sabres were retired on 31 July 1980.

Saudi Arabia

In August and September 1958, 16 ex-USAF F-86Fs modified to F-40 standard were loaned to the 7244th Air Base Group, USAF, at Dhahran, for the training of Saudi pilots. They became a permanent fixture in the country on 11 March 1961, under the terms of MDAP. The Sabres equipped No. 5 Squadron of the Al Quwwat Al-Jawwiya Assa'udiaya (Royal Saudi Air Force), the first RSAF combat squadron. In 1963 and again in 1967, the Sabres were sent to Federal Germany for complete refurbishment. They were supposed to counter the threat posed by Egyptian MiG-19s and MiG-21s, but in this they were completely outclassed and so, in 1966, the RSAF acquired a new air defence system using British Hunters, BAC Strikemasters, and Lightnings. However, during 1966–67 the RSAF acquired more F-86Fs, ex-USAF and ex-Norwegian Sabres, which were delivered as part of MDAP. These Sabres were used to equip No. 7 Squadron, which had become the jet OCU for the RSAF. In December 1969 together with Lightnings operating in the ground-attack role, the RSAF Sabres at Khamis Muchayt, supported by Pakistan Air Force personnel, were used in action against fortified positions on the troubled border with Yemen. When the conflict ended in January 1970, the Sabres returned to Dhahran

and No. 5 Squadron disbanded. Northrop F-5B and F-5E aircraft replaced No. 7 Squadron's surviving Sabres in the period 1971 to 1977.

South Africa

In 1950, No. 2 Squadron, South African Air Force (SAAF) operated F-51D Mustangs alongside the USAF's 18th TFW in Korea. When the SAAF retired its F-51Ds on 31 December 1952, No. 2 Squadron moved from K-10 Chinhae to K-55 Osan to receive the first of 18 F-86F Sabres loaned from the USAF. Its first mission on Sabres was flown on 16 March 1953. By the end of the Korean War on 27 July 1953, No. 2 Squadron's Sabres had flown over 2,000 sorties, losing two F-86Fs to ground fire.

These aircraft are C.5/F-86F Sabres of Ala de Caza (Fighter Wing) 1 of the Ejército del Aire. Ala 1, the first Spanish unit to receive the F-86F, was formed at Manises AB on 6 September 1955, with two escadrones, Nos 11 and 12. Altogether, the EdA received 270 F-86Fs, 125 of them being supplied under MDAP. *(via Robert Jackson)*

With the ending of hostilities, all the Sabres were returned to the USAF and No. 2 Squadron returned to South Africa, where it re-equipped with the de Havilland Vampire Mk 5 and Mk 9. In 1954 South Africa renewed its association with the Sabre when 34 Canadair-built Sabre 6s were ordered, with delivery taking place in August 1956. They equipped No. 1 Squadron at Pietersberg and No. 2 'Cheetah' Squadron at Waterkloof. In June 1963, No. 2 Squadron began re-equipment with the Dassault Mirage IIICZ and its Sabres were re-assigned to No. 1 Squadron. Arms embargoes against South Africa severely hampered the SAAF's re-equipment programme and No. 1 Squadron did not give up its Sabres until 1977, when it re-equipped with the Mirage F1AZ. Twelve of its Sabre 6s were transferred to No. 85 Advanced Flying School at Pietersberg. The Sabres were finally withdrawn on 10 October 1979, but not officially retired until April 1980 when the last of the Sabres were replaced by Mirage IIIs. In January 1981 Flight Systems Inc. at Mojave, California, purchased ten of the retired Sabres.

Spain

Although Spain was not a member of NATO, General Franco's government was very much pro-Western and in return for the use by the United States of three key air bases and a naval base, Spain received substantial quantities of American aid. When, on 26 September 1953, Spain signed a defence agreement with the US, one of the effects was the selection of the F-86F as the first jet fighter to equip the *Ejército del Aire* (EdA, or Spanish air force). During 1955 the EdA was assigned 23 F-86F-25s, or C.5s (Caza, fighter, type 5) as they were called locally. All told, the EdA received 270 F-86Fs, 125 of them being supplied under MDAP. That same year CASA, at its Getafe factory, began the conversion of these Sabres to F-40 standard. The first 13 refurbished F-86Fs were handed over to the EdA at Getafe on 6 October 1956. By the end of January 1957, 152 Sabres were in service, equipping three Alas de Caza (Fighter Wings) totalling nine escadrones (squadrons). Ala 1 formed at Manises AB on 6 September 1955 and

was followed by Ala 2 at Zaragosa and Ala 4 at Son San Juan on 19 September 1956. In 1957, 98 Escadrón was established at Torrejón to operate 13 Sabres and helicopters, the first jets being assigned on 26 June 1959. In April 1985, Escadrón 98 was renamed 981 Escadrón. Two more Alas de Caza had formed in 1959 – Ala 5 at Morón on 8 May and Ala 6 at Torrejón on 6 June. Beginning in February 1967, the Sabres were progressively withdrawn from service and by February 1971, 102 Escadrón at Torrejón was the only F-86F/C.5 Sabre unit remaining in the EdA inventory. The last EdA Sabre flight took place on 7 December 1972 and the aircraft was officially retired on 31 December.

Thailand

Early in 1960 the Royal Thai Air Force (RTAF) received 20 F-86F-30 Sabres which were issued to No. 12 Squadron of the 1st Wing at Don Muang and beginning the re-equipment of No. 13 Squadron in November that year, replacing the F-84G Thunderjet and F8F-1 Bearcat, respectively. On 7 March 1962, a second batch of 26 ex-USAF F-86Fs was delivered to Thailand. Like the F-86F-30s, these too were upgraded to F-86-40 standard. Including attrition replacements, the RTAF finally gained about 47 Sabres, which were known in service as the B.Kh.17 (Boh Khoh, aircraft, fighter). In 1964, 20 ex-US ANG (mainly ex-124th FIS, Iowa ANG) F-86L Sabres were acquired. Known in Thai service as the B.Kh.17K, these were assigned to No. 12 Squadron to provide a limited all-weather interception capability. When No. 12 Squadron converted to the counter-insurgency role with the OV-10 Bronco, the Squadron's F-86Fs were re-assigned to form No. 43 Squadron of the 4th Wing, which was activated at Takhli AB. In April 1966, No. 13 Squadron began re-equipping with Northrop F-5As and its F-86Fs were also transferred to No. 43 Squadron, which operated F-86F Sabres until 1972 when they were replaced by the Cessna A-37B at Nakhon Phanom. Northrop F-5Es finally replaced the F-86L Sabres in 1976.

Tunisia

In 1969 the Tunisian Republican Air Force received from the USA 12 ex-JASDF F-86F-40 Sabres, which were in storage at Davis-Monthan AFB. They were used to equip No. 11 Squadron at Sidi Ahmed AB, near Bizerte, but attrition substantially reduced this small force and the survivors were used mainly for training until 1984, when Italian-built MB.326K aircraft replaced them.

Turkey

In June 1954, the first 34 of 105 Canadair Sabre Mk 2s and Mk 4s (refurbished to F-86(M) standard) were delivered to the *Turk Hava Kuvvetleri* (THK, or Turkish Air Force). They were used to equip three filos (squadrons) at 4th Hava Ussu (Air Base) at Eskisehir. First to take delivery was 141 Filo on 7 September 1954 with 142 Filo following on 1 February 1955, and 143 Filo that summer. In 1956, all of Turkey's Sabres were moved to Merzifon as part of the 44th Ucus Grubo (Fighter-Bomber Group). Some F-86Es were also delivered from the NAA production line to make up for attrition. In August 1964, 141 Filo converted to the F-104G Starfighter and 142 Filo changed to the F-5A in 1967. A year later 143 Filo also converted to the F-5A Freedom Fighter.

Venezuela

In 1955, the *Fuerza Aérea Venezolanas* (FAV, or Venezuelan Air Force) asked the USA for Sabres to replace its obsolete de Havilland Vampire fighters. By 1957, Venezuela had received 22 MDAP-funded F-86Fs and these were operated by Escuadrón de Caza 36 'Jaguares' of Grupo Aérea de Caza 12 at Mariscal Sucre. In the mid-1960s the Venezuelan government placed orders with the West German government for 78 ex-Luftwaffe F-86K Sabres. Those that entered service replaced the de Havilland Vampire FB.Mk 5 in Escuadrones de Caza-Bombardeo 34 and 35. Later, F-86Ks replaced Escuadrón de Caza 36's F-86Fs. The Dassault Mirage IIIV finally replaced the last of the F-86Ks in 1971.

West Germany

Sabres served with the *Luftwaffe* between 1958 and 1964. In 1957, the Canadian government generously presented 75 CL-13A Sabre 5s (formerly operated by the RCAF in Europe) to the German Federal Republic. The Mk 5 was used by Waffenschule-10 (WS10) at Oldenburg for converting pilots to the Sabre 6 day-interceptor and the Fiat-built F-86k all-weather fighter (of which 88 were used to equip Jagdgeschwader 74 at Neuburg). On 9 October 1958, Canadair produced the last CL-13B (F-86E) Sabre 6. Some 225 Sabre 6s were supplied to the German Federal Republic. The Sabre 6 was delivered to Waffenschule-50 and the majority served in the air-superiority role with three operational day fighter wings (Jagdgeschwader 71, 72, and 73), and some with Erprobungsstelle 61. In 1964 JG71, which on 21 April 1961 had become JG71 'Richthofen', replaced its Sabres with the F-104G Starfighter. In October 1964, JG72 and JG73 were redesignated Jagdbombergeschwader (JaBoG) and were renamed JaBoG 43 and JaboG 42, respectively. Starting in April 1966, JaboG 42 and JaboG 43 began conversion to the Fiat G.91R. Condor Flugdienst operated seven Sabres 6s modified for target towing from 1 October 1966 to April 1974.

Yugoslavia

In the mid-1950s, MDAP donated to the Yugoslav air force several ex-RAF Canadair Sabre Mk 4s refurbished to F-86E(M) standard in the UK. Yugoslavia was to receive 79 Sabres, but deliveries ended after just 43 F-86E(M) aircraft had been received when, on 15 July 1957, the Yugoslav government, which by now had moved closer to the Soviet Bloc, announced that it would no longer accept any further American military aid from December that year. However, in the early 1960s the Yugoslav government then purchased 78 ex-RAF F-86E(M)s, and 130 F-86D all-weather interceptors. The last remaining F-86Es were withdrawn from service in 1971 and the F-86Ds finally disappeared in late 1974.

Appendix 1. Weapons

The F-86A Sabre's standard armament was six 0.5-in (12.7-mm) Browning machine-guns, each fed by a belt containing semi-armour-piercing/incendiary, tracer and HE rounds. Typically, loading was AP and incendiary with every 5th round tracer. The rounds weighed only 48 g (1.69 oz) but muzzle velocity was high. Because the MiG-15 was initially intended as a heavy-bomber destroyer, its standard armament was two NR-23 cannon (with 80 rounds per gun) and one N-37 cannon (with 40 rounds). Soviet designers estimated that on average only two 37-mm or eight 23-mm shell hits would be enough to destroy a B-29. Trying to bring down a fast, highly manoeuvrable fighter when the range was constantly changing, was, however, an entirely different matter. Both the MiG's guns had the same muzzle velocities (2,264 ft per second), but entirely different ballistic qualities, so they dropped at vastly different trajectories. (The 23-mm shell weighed 0.20 kg (0.44 lb) while the 37-mm shell weighed 0.73 kg (1.62 lb). This made harmonisation of the guns and sight almost impossible and sighting was a real problem at all but very close range. The MiG-15's slower rate of fire (the 23-mm cannon's rate of fire was 14 rounds per second; the 37-mm weapon fired 7 rounds per second) and different gun trajectories could result in a short two-second burst having a widely spread pattern, with relatively few rounds hitting their target. (On occasion F-86 pilots under attack by MiGs noted 23-mm rounds passing above them while the 37-mm shells went

below.) In a two-second burst the Sabre pilot could hit the target with a much greater number of rounds, and alI in a very tight grouping. The F-86's radar ranging and gyroscopic sight made for greater accuracy (the MiG-15bis had a simple optical ASP-3N sight) and the lighter .50-cal rounds had near ideal ballistic qualities, with greater accuracy at extreme ranges. Normally, the guns were harmonised so that the six trajectories converged 1,000 ft (305 m) ahead of the F-86. The six Brownings had a rate of fire of 1,200 rounds per minute, or 13 seconds of firing for the normal capacity of 267 rounds per gun. Even so, Sabre pilots could literally pump several hundred rounds into a MiG without it going down.

Experience in Korea resulted in the *Gun-Val* project after requests from pilots for heavier calibre guns, while still retaining most of the rate of fire of the .50-cal guns. Four F-86E-10s and six F-86F-ls were armed with four T-160 20-mm guns and re-designated F-86F-2s, while two F-86F-ls armed with Oerlikon-type 20-mm guns were redesignated F-86F-3s. Eight F-86F-2s were issued to the 4th FIW in March 1953, for 16 weeks of combat tests. Although only two of the guns could be safely fired at once because of engine compressor stalling, the *Gun-Vals* proved successful. Of 41 MiGs fired upon, six were destroyed, three were probably destroyed, and 13 damaged. Belt fed, and firing 1,500 rounds per minute, the T-160 went into production as the M-39 and was used on most F-86H Sabres and all F-100 series fighters.

F-86D-15 (NA-165) 50-574 is shown here with its retractable ventral pack lowered and firing a battery of 24 2.75-in (70-mm) Mighty Mouse 18-lb (8.16-kg) folding-fin rockets. An electronic link existed between the search radar and the AN/APA-84 computer, which calculated the pilot's lead-collision course, (the pilot following it by keeping the 'blip' on the AI radar scope inside a 1-in (2.54-cm) circle). When the automatic tracking system indicated 20 seconds to target, the system would instruct the pilot to turn to a 90° collision course. At this point he could launch six, 12 or all 24 Mighty Mouse rockets. The computer controlled the actual firing, extending the rocket pack in half a second and initiating the firing sequence when the target was about 500 yards (457 m) distant. All 24 rockets could be fired in ⅛ second. The rocket pack retracted in just over 3 seconds and a symbol on the radar scope, which illuminated at a range of 250 yards (229 m), warned the pilot to break off the attack. In 1953, the need to pass information from the ground radar warning system to the interceptor force by voice communication was eliminated with the implementation of the SAGE (Semi-Automatic Ground Environment) system developed by the Massachusetts Institute of Technology. This early data link system transmitted target information to a receiver in the interceptor's cockpit and automatically positioned the fighter for a lead-collision attack using the E-4 fire control system. *(NAA via Jerry Scutts)*

Appendix 2. Production Details

Model	Serial number	Quantity	Model	Serial number	Quantity
XP-86 (NA-140)	45-59597/59599	3	F-86A-5 (NA-161)	49-1007/1339	333
F-86A-1 (NA-151)	47-605/637	33	YF-86D (NA-164)	50-577/578	2
F-86A-5 (NA-151)	48-129/316	188	F-86D-1 (NA-165)	50-455/491	37

Model	Serial number	Quantity
F-86D-5 (NA-165)	50-492/517	26
F-86D-10 (NA-165)	50-518/553	36
F-86D-15 (NA-165)	50-554/576 50-704/734	54
F-86D-20 (NA-177)	51-2944/3131	188
F-86D-25 (NA-173)	51-5857/5944	88
F-86D-30 (NA-173)	51-5945/6144	200
F-86D-35 (NA-173)	51-6145/6262 51-8274/8505	350
F-86D-40 (NA-190)	52-3598/3897	300
F-86D-45 (NA-190)	52-3898/4197	300
F-86D-50 (NA-190)	52-4198/4304	301
F-86D-55 (NA-201)	53-557/781	225
F-86D-60 (NA-201)	53-782/1071 53-3675/3710 53-4018/4090	399
F-86E-1 (NA-170)	50-579/638	60
F-86E-5 (NA-170)	50-639/689	51
F-86E-6 (CL-13 Mk 2)	52-2833/2892	60
F-86E-10 (NA-172)	51-2718/2849	132
F-86E-15 (NA-172)	51-12977/13069	93
F-86F-1 (NA-172)	51-2850/2927	78
F-86F-5 (NA-172)	51-2928/2943	16
F-86F-10 (NA-172)	51-12936/12969	34
F-86F-15 (NA-172)	51-12970/12976	7
F-86F-20 (NA-176)	51-13070/13169	100
F-86F-25 (NA-176)	51-13170/13510	341
F-86F-30 (NA-191)	52-4305/5163	859
F-86F-35 (NA-191)	52-5164/5271	108
F-86F-25 (NA-193)	52-5272/5530	259
F-86F-35 (NA-202)	53-1072/1228	157
F-86F-40 (NA-227)	55-3816/4030 55-4983/5047	280
F-86F-40 (NA-231)/Mitsubishi	55-5048/5117	70
F-86F-40 (NA-238)/Mitsubishi	56-2773/2882	110
F-86F-40 (NA-256)/Mitsubishi	57-6338/6457	120
F-86H-1 (NA-187)	52-1975/1976	2
F-86H-1 (NA-187)	52-1977/2089	113

Model	Serial number	Quantity
F-86H-5 (NA-187)	52-2090/2124 52-5729/5753	60
F-86H-10 (NA-203)	53-1229/1528	300
YF-86K (F-86D-40) (NA-205)	52-3630 and 52-3804	2
F-86K (NA-213)	54-1231/1350	120
F-86K (NA-207)/Fiat	53-8273/8322	50
F-86K (NA-221)/Fiat	55-4811/4880	70
F-86K (NA-232)/Fiat	55-4881/4936	56
F-86K (NA-242)/Fiat	56-4116/4160	45
XFJ-1 (NA-134)	39053/39055	3
FJ-1 (NA-141)	120342/120371	30
XFJ-2B (NA-185)	133754/133756	3
FJ-2 (NA-181)	131927/132126	300
FJ-3/FJ-3M (NA-194)	135774/136162	389
FJ-3/FJ-3M (NA-215)	139210/139278 141364/141443	149
FJ-4 (NA-208)	139279 and 139280	2
FJ-4 (NA-209)	139281/139323 139424/139530	150
FJ-4B (NA-209)	139531/139555	25
FJ-4B (NA-209)	141444/141489	46
FJ-4B (NA-244)	143493/143643	184
CL-13 Sabre Mk 1	19101	1
CL-13 Sabre Mk 2	19102/19199 19201/19452	350
CL-13 Sabre Mk 3	19200	1
CL-13 Sabre Mk 4	19453/19890	438
CL-13A Sabre Mk 5	23001/23370	370
CL-13B Sabre Mk 6	2021/2026 for Columbia	6
CL-13B Sabre Mk 6	350/383 for South Africa	34
CL-13B Sabre Mk 6	23371/23662	292
CL-13B Sabre Mk 6	23663/23760	90
CL-13B Sabre Mk 6	for Luftwaffe	225
CA-26	A94-101	1
CA-27 Sabre Mk 30	A94-901/921	21
CA-27 Sabre Mk 31	A94-922/942	21
CA-27 Sabre Mk 32	A94-943/990 A94-351/371	69

CL-13 Sabre 6s rest at Prestwick on 11 May 1963 en route home to Canada from Germany. The last RCAF Sabres in Europe were retired on 14 November 1963, when No. 439 Squadron at Marville, France, was stood down. (Harry Holmes)

Appendix 3. Specifications

Parameter	F-86A	F-86E	F-86F	F-86H	F-86D
Dry thrust	23.12-kN (5,200-lb st)	23.12-kN (5,200-lb st)	26.55-kN (5,970-lb st)	39.67-kN (8,920-lb st)	25.35-kN (5,700-lb st)
Afterburning thrust					33.93-kN (7,630-lb st)
Engine	J47-GE-13	J47-GE-13	J47-GE-27	J73-GE-3E	J47-GE-17
Wing span	11.30 m (37 ft 1 in)	11.30 m (37 ft 1 in)	11.30 m (37 ft 1 in)	11.91 m (39 ft 1 in)	11.30 m (37 ft 1 in)
Length	11.43 m (37 ft 6 in)	11.43 m (37 ft 6 in)	11.43 m (37 ft 6 in)	11.79 m (38 ft 8 in)	12.29 m (40 ft 4 in)
Height	4.50 m (14 ft 9 in)	4.50 m (14 ft 9 in)	4.50 m (14 ft 9 in)	4.57 m (15 ft)	4.57 m (15 ft)
Wing area	26.76 m² (288 sq ft)	26.76 m² (288 sq ft)	26.76 m² (288 sq ft)	29.08 m² (313 sq ft)	26.76 m² (288 sq ft)
Empty weight	4761 kg (10,495 lb)	4919 kg (10,845 lb)	4967 kg (10,950 lb)	6276 kg (13,836 lb)	5656 kg (12,470 lb)
Gross weight	7420 kg (16,357 lb)	8077 kg (17,806 lb)	7711 kg (17,000 lb)	9912 kg (21,852 lb)	7757 kg (17,100 lb)
Maximum speed at sea level	1093 km/h 679 mph	1093 km/h 679 mph	1118 km/h 695 mph	1114 km/h 692 mph	1114 km/h 692 mph
Maximum speed (speed/ at altitude)	967 km/h/ 10668 m (601 mph/ 35,000 ft)	967 km/h/ 10668 m (601 mph/ 35,000 ft)	978 km/h/ 10668 m (608 mph/ 35,000 ft)	993 km/h/ 10668 m (617 mph/ 35,000 ft)	985 km/h/ 12190 m (612 mph/ 40,000 ft)
Cruising speed	848 km/h (527 mph)		837 km/h (520 mph)	970 km/h (603 mph)	
Initial climb rate	37.95 m/s (7,470 ft/min)	36.83 m/s (7,250 ft/min)	50.80 m/s (10,000 ft/min)	65.53 m/s (12,900 ft/min)	90.42 m/s (17,800 ft/min)
Service ceiling	14630 m (48,000 ft)	14387 m (47,200 ft)	15240 m (50,000 ft)	14935 m (49,000 ft)	16642 m (54,600 ft)
Range	1263 km (785 miles)		2044 km (1,270 miles)	1674 km (1,040 miles)	1345 km (836 miles)
Armament	6 x 0.50-in (12.7-mm) machine-guns, plus 2 x 1,000-lb (454-kg) bombs, or 16 x 5-in (127-mm) rockets	6 x 0.50-in (12.7-mm) machine-guns, plus 2 x 1,000-lb (454-kg) bombs, or 16 x 5-in (127-mm) rockets	6 x 0.50-in (12.7-mm) machine-guns, plus 2 x 1,000-lb (454-kg) bombs, or 16 x 5-in (127-mm) rockets	4 x 20-mm cannon, plus 2 x 1,000-lb (454-kg) bombs, or 16 x 5-in (127-mm) rockets	24 x 2.75-in (70-mm) FFAR

Parameter	FJ-2 Fury	FJ-3/-3M Fury	FJ-4 Fury	CL-13 SABRE 2/4	CL-13B SABRE 6
Dry thrust	26.68-kN (6,000-lb st)	34.02-kN (7,650-lb st)	25.35-kN (5,700-lb st)	23.12-kN (5,200-lb st)	32.35-kN (7,275-lb st)
Afterburning thrust			33.93-kN (7,630-lb st)		
Engine	J47-GE-2	J65-W-4B	J65-W-16A	J47-GE-13	Orenda 14
Wing span	11.30 m (37 ft 1 in)	11.30 m (37 ft 1 in)	11.91 m (39 ft 1 in)	11.56 m (37 ft 11 in)	11.56 m (37 ft 11 in)
Length	11.46 m (37 ft 7 in)	11.43 m (37 ft 6 in)	11.07 m (36 ft 4 in)	11.43 m (37 ft 6 in)	11.43 m (37 ft 6 in)
Height	4.14 m (13 ft 7 in)	4.17 m (13 ft 8 in)	4.24 m (13 ft 11 in)	4.50 m (14 ft 9 in)	4.50 m (14 ft 9 in)
Wing area	26.75 m^2 (287.90 sq ft)	28.08 m^2 (302.30 sq ft)	31.46 m^2 (338.66 sq ft)	26.75 m^2 (287.90 sq ft)	28.08 m^2 (302.30 sq ft)
Empty weight	5353 kg (11,802 lb)	5536 kg (12,205 lb)	5992 kg (13,210 lb)	4733 kg (10,434 lb)	4816 kg (10,618 lb)
Gross weight (loaded weight for Canadair Sabres)	8565 kg (18,882 lb)	8131 kg (17,926 lb)	8094 kg (17,845 lb)	6612 kg (14,577 lb)	6628 kg (14,613 lb)
Maximum speed (speed/ at altitude)	969 km/h/ 10668 m (602 mph/ 35,000 ft)	1003 km/h/ 10668 m (623 mph/ 35,000 ft)	1015 km/h/ 7620 m (631 mph/ 25,000 ft)	949 km/h (590 mph) at sea level	975 km/h (606 mph) at sea level
Cruising speed (speed/ at altitude)	970 km/h/ 13381 m (603 mph/ 43,900 ft)	846 km/h/ 13198 m (526 mph/ 43,300 ft)	859 km/h/ 13929 m (534 mph/ 45,700 ft)		787 km/h/ 13716 m (489 mph/ 45,000 ft)
Climb to altitude	9144 m (30,000ft) 11 minutes		9144 m (30,000ft) 6 minutes 18 seconds	10668 m (35,000ft) 8 minutes 12 seconds	10668 m (35,000ft) 4 minutes 8 seconds
Initial climb rate		42.93 m/s (8,450 ft/min)			59.94 m/s (11,800 ft/min)
Service ceiling	12710 m (41,700 ft)	14935 m (49,000 ft)	14265 m (46,800 ft)	14387 m (47,200 ft)	16459 m (54,000 ft)
Range	1593 km (990 miles)	1593 km (990 miles)	2390 km (1,485 miles)		
Armament	4 x 20-mm cannon	4 x 20-mm cannon, plus 2 x AAM-N-7/GAR-8 Sidewinders or 2 x 1,000-lb (454-kg) bombs, 2 x 500-lb (227-kg) bombs or 4 rocket packs	4 x 20-mm cannon, plus 2 x AAM-N-7/ GAR-8 Sidewinders or 2 x 1,000-lb (454-kg) bombs, 2 x 500-lb (227-kg) bombs or 4 rocket packs, or 5 x Bullpup ASMs plus guidance pod, or 1 x Mk 7, Mk 12, or Mk 28 special store	6 x 0.50-in (12.7-mm) machine-guns, plus 2 x 1,000-lb (454-kg) bombs, or 16 x 5-in (127-mm) rockets	6 x 0.50-in (12.7-mm) machine-guns, plus 2 x 1,000-lb (454-kg) bombs, or 16 x 5-in (127-mm) rockets

Appendix 4. Model Kits

Academy

Aircraft	Scale
F-86F Mike's Bird	1:72
F-86E	1:72
F-86F-30	1:48
F-86F MiG Killer	1:48

Airfix

Aircraft	Scale
F-86F	1:72

Fujimi

Aircraft	Scale
F-86 (silver plated)	1:72
F-86F JASDF	1:72
F-86F USAF	1:72
RF-86F JASDF	1:72
F-86F 'Blue Impulse'	1:72

Hasegawa

Aircraft	Scale
F-86F	1:72
F-86D JASDF	1:72
F-86D USAF	1:72
F-86D USAF/Far East	1:72
F-86/RF-86 JASDF	1:72
QF-86F Sabre Drone	1:48
F-86F-30 USAF	1:48
F-86F-40 JASDF	1:48
F-86F-40 'Blue Impulse'	1:48
F-86F The Huff	1:48
F-86F-40 early Japanese	1:48
F-86F-35 Sky Blazers	1:48
F-86F-40 JASDF	1:32
F-86F	1:32

Heller

Aircraft	Scale
F-86F Sabre CL-13B	1:72
F-86F	1:72

Hobbycraft

Aircraft	Scale
F-86F-25/F-30	1:72
F-86E	1:72
F-86 USMC	1:72
Canadair Sabre 5	1:72
F-86F-25	1:72
Canadair Sabre Mk 6	1:72

PM Model

Aircraft	Scale
F-86E	1:72

Revell

Aircraft	Scale
F-86D	1:48

Information on the model kits was supplied by H. G. Hannant Ltd: www.hannants.co.uk

Appendix 5. Further Reading

F-86 Sabre: The Operational Record by Robert Jackson, Airlife, 1994

North American F-86 Sabre by Duncan Curtis, Crowood Aviation Series, 2000

The North American Sabre by Roy Wagner, MacDonald, London, 1963

Warbirds Directory by John Chapman and Geoff Goodall, Ed. Paul Coggan, Warbirds Worldwide

F-86 Sabre In Action by Larry Davis, Squadron/Signal Publications Inc. 1992

The Engagement by Reg Adams, Sabre Jet Classics Vol. 9. No. 1, Winter 2001

War In Peace: An Analysis of Warfare from 1945 to the present day, Black Cat, 1988

The Royal Air Force At War by Martin W. Bowman, PSL, 1997

Air War over Korea by Robert Jackson, Ian Allan Ltd, 1973

Air War Korea 1950–1953 by Robert Jackson, Airlife, 1998

Battle for Pakistan by John Fricker, Ian Allan, London, 1978

Pakistan's Sabre Ace by Jon Guttman, Air History, September 1998

Eye-witness to M. M. Alam's encounter with the IAF by M. A. Iqbal, PIADS, 1999

Sqn Ldr Sarfaraz Ahmed Rafiqui Kaiser Tufail, Defence Journal, September 1998

Pakistan Fiza'ya: Psyche of the Pakistan Air Force by Pushpinder Singh Chopra, Ravi Rikhye and Peter Steinemann, Society for Aerospace Studies, New Delhi, 1991, distributed by Himalayan Books

Official Site of the Pakistan Air Force Museum

Laying the Sargodha Ghost to Rest by Pushpinder Singh Chopra, Vayu Aerospace Review, November 1985

My Years with the IAF by Air Chief Marshal (retd.) Pratap Chandra Lal, Lancer International, New Delhi, 1986

Canadian Aircraft Since 1909 by K. M. Molson and H. A. Taylor, Putnam, 1982

The Canadair Sabre by Larry Milberry, Canav Books, 1986

The Squadrons of the Royal Air Force and Commonwealth 1918–1988 by James J. Halley, Air-Britain, 1988

British Military Aircraft Serials and Markings, BARG, Hollen Street Press, 1980

Stars & Bars: A Tribute to the American Fighter Ace 1920–1973 by Frank Olynyk, Grub Street, 1995

The United States Air National Guard by René J. Francillon, Aerospace Publishing Ltd, 1993

Korea: The Air War 1950–1953 by Jack C. Nicholls and Warren E. Thompson, Osprey, 1991

South to the Naktong, North to the Yalu by R. E. Appleman, US Department of the Army, Washington, 1961

The United States Air Force in Korea 1950–53 by Robert F. Futrell, Duell, Sloan and Pearce, New York, 1961

Five Down and Glory by Gene Gurney and P. Friedlander, Putnam, New York, 1958

Korea 1951–1953 by John Miller Jr., Owen J. Carroll and Margaret E. Tackley, Office of the Chief of Military History Washington, 1956

Korea – the Limited War by D. Rees, MacMillan, London, 1964

Airpower: the Decisive Force in Korea by Col James T. Stewart, Van Nostrand, 1957

Air Combat 'MiG Maulers' F-86 In Korea, 1950–53 by Larry Davis with Warren Thompson, Wings of Fame Vol. 11, Aerospace Publishing Ltd, 1998

F-86F-30 3335 of the Fuerza Aerea Venezolanas (FAV or Venezuelan air force). By 1957 Venezuela had received 22 MDAP-funded F-86Fs and these were operated by Escuadrón de Caza 36 'Jaguares' of Grupo Aerea de Caza 12 at Mariscal Sucre. *(via Robert Jackson)*

Index